The
Environment
OPPOSING VIEWPOINTS®

Other Books of Related Interest

Opposing Viewpoints Series

Africa
America Beyond 2001
America's Cities
Animal Rights
Endangered Species
Global Resources
Population
The Third World
21st Century Earth
Water

Current Controversies Series

Energy Alternatives
Hunger
Pollution

At Issue Series

Environmental Justice

The Environment

OPPOSING VIEWPOINTS®

David Bender & Bruno Leone, *Series Editors*

A.E. Sadler, *Book Editor*

OPPOSING
VIEWPOINTS®
SERIES

Greenhaven Press, Inc., San Diego, CA

Photo: Gazelle Technologies

Greenhaven Press, Inc.
PO Box 289009
San Diego, CA 92198-9009

Library of Congress Cataloging-in-Publication Data

The environment : opposing viewpoints / A.E. Sadler, book editor.
    p.   cm. — (Opposing viewpoints series)
  Includes bibliographical references and index.
  ISBN 1-56510-397-1 (lib. : alk. paper). —
ISBN 1-56510-396-3 (pbk. bdg. : alk. paper)
  1. Environmental degradation. 2. Sustainable development.
3. Human Ecology—United States. I. Sadler, A.E., 1961–
II. Series: Opposing viewpoints series (Unnumbered)
GE140.E554 1996
363.7—dc20   95-51743
                                                      CIP

"Congress shall make no law . . .
abridging the freedom of speech,
or of the press."

First Amendment to the U.S. Constitution

The basic foundation of our democracy is the First Amendment
guarantee of freedom of expression. The Opposing Viewpoints
Series is dedicated to the concept of this basic freedom and the
idea that it is more important to practice it than to enshrine it.

Contents

Chapter 4: Is Ecological Conservation Bad for the Economy?

Chapter 5: How Should the Environment Be Protected?

Why Consider Opposing Viewpoints?

"The only way in which a human being can make some approach to knowing the whole of a subject is by hearing what can be said about it by persons of every variety of opinion and studying all modes in which it can be looked at by every character of mind. No wise man ever acquired his wisdom in any mode but this."

John Stuart Mill

In our media-intensive culture it is not difficult to find differing opinions. Thousands of newspapers and magazines and dozens of radio and television talk shows resound with differing points of view. The difficulty lies in deciding which opinion to agree with and which "experts" seem the most credible. The more inundated we become with differing opinions and claims, the more essential it is to hone critical reading and thinking skills to evaluate these ideas. Opposing Viewpoints books address this problem directly by presenting stimulating debates that can be used to enhance and teach these skills. The varied opinions contained in each book examine many different aspects of a single issue. While examining these conveniently edited opposing views, readers can develop critical thinking skills such as the ability to compare and contrast authors' credibility, facts, argumentation styles, use of persuasive techniques, and other stylistic tools. In short, the Opposing Viewpoints Series is an ideal way to attain the higher-level thinking and reading skills so essential in a culture of diverse and contradictory opinions.

In addition to providing a tool for critical thinking, Opposing Viewpoints books challenge readers to question their own strongly held opinions and assumptions. Most people form their opinions on the basis of upbringing, peer pressure, and personal, cultural, or professional bias. By reading carefully balanced opposing views, readers must directly confront new ideas as well as the opinions of those with whom they disagree. This is not to simplistically argue that everyone who reads opposing views will—or should—change his or her opinion. Instead, the series enhances readers' depth of understanding of their own views by encouraging confrontation with opposing ideas. Careful examination of others' views can lead to the readers' understanding of the logical inconsistencies in their own opinions, perspective on why they hold an opinion, and the consideration of the possibility that their opinion requires further evaluation.

Evaluating Other Opinions

To ensure that this type of examination occurs, Opposing Viewpoints books present all types of opinions. Prominent spokespeople on different sides of each issue as well as well-known professionals from many disciplines challenge the reader. An additional goal of the series is to provide a forum for other, less known, or even unpopular viewpoints. The opinion of an ordinary person who has had to make the decision to cut off life support from a terminally ill relative, for example, may be just as valuable and provide just as much insight as a medical ethicist's professional opinion. The editors have two additional purposes in including these less known views. One, the editors encourage readers to respect others' opinions—even when not enhanced by professional credibility. It is only by reading or listening to and objectively evaluating others' ideas that one can determine whether they are worthy of consideration. Two, the inclusion of such viewpoints encourages the important critical thinking skill of objectively evaluating an author's credentials and bias. This evaluation will illuminate an author's reasons for taking a particular stance on an issue and will aid in readers' evaluation of the author's ideas.

As series editors of the Opposing Viewpoints Series, it is our hope that these books will give readers a deeper understanding of the issues debated and an appreciation of the complexity of even seemingly simple issues when good and honest people disagree. This awareness is particularly important in a democratic society such as ours in which people enter into public debate to determine the common good. Those with whom one disagrees should not be regarded as enemies but rather as people whose views deserve careful examination and may shed light on one's own.

Thomas Jefferson once said that "difference of opinion leads to inquiry, and inquiry to truth." Jefferson, a broadly educated man, argued that "if a nation expects to be ignorant and free . . . it expects what never was and never will be." As individuals and as a nation, it is imperative that we consider the opinions of others and examine them with skill and discernment. The Opposing Viewpoints Series is intended to help readers achieve this goal.

David L. Bender & Bruno Leone,
Series Editors

Introduction

"Has the balance shifted, even if subtly, such that groups saying 'We must protect the environment' find a more skeptical audience now than they did a decade or two decades ago?"

Richard Morin, Washington Post National Weekly Edition,
June 5–11, 1995

The first Earth Day, held on April 22, 1970, marked the beginning of the modern environmental movement, today recognized as one of the most influential social movements in U.S. history. In its first decade, the environmental movement propelled through the federal legislature a series of laws designed to protect water and air quality, wilderness areas, and endangered species. By the middle of its third decade, the movement had grown so popular that the vast majority of Americans were identifying themselves as "environmentalists." A 1995 study published by the American Enterprise Institute, a politically conservative think tank, reported that "the environment has now become a 'core value' to most Americans." Everyone, it seemed, had gone "green."

Ironic as it might sound, widespread acceptance nearly brought the environmental movement to a halt during the early to mid-1990s. As the movement expanded to encompass an ever-broadening population of supporters, environmentalism as an idea began to grow diffuse and to lose much of the meaning it had held among its early followers. Confusing the situation further, the industrial sector—long recognized as the foe against whom environmentalists had rallied—began publicly addressing environmental issues. As some companies began informing consumers of their efforts to improve the environment, the traditional chasm between industry and environmentalists grew less distinct. While some national environmental organizations welcomed industrial newcomers to the environmental movement, many activists—especially at the local level—viewed them with suspicion. Critics charged that many in the business sector had joined the environmental movement as a means of improving their image and furthering their own interests. And many criticized the national environmental organizations for being manipulated by, and willing to compromise with, these corporate strategists. "Rather than beat environmentalists," wrote Christopher Hartlove, "industry has claimed to join them," a strategy he believed had left the movement "flat-footed."

Other activists believed that getting stalled was simply part of the natural cycle of activism. "You can see this type of thing happening in a lot of social movements," said Eric Olson, senior attorney for the Natural Resources Defense Council. "In the beginning . . . the problems are seen as easy to fix, and some solutions come quickly." It was true that the most tangible problems, such as polluted skies and waterways, while not eradicated, had at least been lessened. Even its critics seldom denied that the environmental movement had made America cleaner and healthier. Yet according to Olson, with the more visible issues now addressed, the problems remaining—like global warming and ozone depletion—were those that proved most obscure and difficult to convey to a mass audience. "The problems get more complex," Olson says, asserting that this leads to a lapse in momentum that leaves the movement vulnerable to its detractors. "The opposition steps in. That's where we are now."

Enter the Wise Use movement, a loose coalition of western ranchers, oil companies, and logging and mining interests who argue for the commercial development and private ownership of federal lands and against government regulation of private lands. Their arguments are not new; they have been aired in America for one hundred years or more. Before the emergence of the Wise Use movement in the 1990s, they were expounded by the Sagebrush Rebellion nearly ten years earlier. The Sagebrush Rebellion is credited with helping to put Ronald Reagan in the Oval Office and tightly allying itself with James Watt, whom Reagan appointed to head the Department of the Interior and who spoke out in favor of privatizing government land.

As its successor, the Wise Use movement has proven an equally formidable force. By 1995, the movement counted more than 1000 member organizations among its ranks and wielded considerable political clout. Its leaders, Ron Arnold and Alan Gottlieb, who run the Council for the Defense of Free Enterprise, based in Seattle, Washington, have expressed strong libertarian leanings and support for gun ownership rights along with their staunch defense of private property rights. Like environmentalists, the Wise Use movement stresses the vital link between human welfare and the ecology of the planet. But unlike the environmentalists, who advocate environmental protection through government regulation, Wise Users advocate private stewardship of the land. They argue that because their interests are tied to their property, individual landowners have a greater incentive than the government does for proper maintenance and conservation. "If you look down a neighborhood street," Wise

Use supporter Nancie Marzulla offers, "you can usually tell which houses are occupied by owners and which is the rental house."

Wise Users see their movement as a populist, grassroots form of activism, and they see themselves, according to journalist Russell Shaw, as the "true environmentalists." They deem their members to be among those who live most closely to the land and who labor exhaustively to draw subsistence from it. In fact, many activists argue that the war is not between environmentalists and traditional polluters, but between those who dwell in cities and those who live in the countryside. Many people believe that urban environmentalists, concerned only with preserving wilderness as a vacation spot, are indifferent to rural residents whose economic survival depends on utilizing their region's natural resources. "We have this novel idea that people, too, are part of the environment," says Chuck Cushman, whose National Inholders Association represents owners of property inside and/or bordering national parklands.

Environmentalists reject Wise Users' argument that private stewardship is best for the environment. "I'd be hard-pressed to provide an instance where private ownership . . . has been used for anything other than commercial benefit," says Deborah Rephan of Greenpeace. "They ought really to be called the 'Me First!' movement," write Stewart L. Udall, who serves as chairman of the Archeological Conservancy, and W. Kent Olson, former executive director of the Nature Conservancy of Connecticut. Udall and Olson argue that the Wise Use agenda, which includes mining and oil drilling in national parks as well as the dismantling of federal air and water quality regulations, embodies "a passion for wrecking America's public lands and environmental laws."

Wise Users are equally critical of environmentalists. "They think they are the only ones with the knowledge to make the rules," says Clark Collins, director of the Blue Ribbon Coalition, which represents some 3000 off-roading enthusiasts. "They are trying to represent themselves as the environmental conscience of the country, but they are not."

The Environment: Opposing Viewpoints presents the arguments introduced by both groups in response to several key questions that confront humankind—Is There an Environmental Crisis? How Serious Is Air and Water Pollution? Is the American Lifestyle Bad for the Environment? Is Ecological Conservation Bad for the Economy? How Should the Environment Be Protected? Throughout this anthology, authors discuss the status of the environment in America and around the globe.

Is There an Environmental Crisis?

The Environment

Chapter Preface

In November 1994, *Fortune* magazine reported that environmentalists were "on the run." According to the magazine, public officials, business leaders, and local citizens alike had grown fed up with the "silly science" of gloom and doom frequently promulgated by environmental activists.

The preceding year had witnessed an increase in books and media coverage that challenged the credibility of claims portending ecological disaster. Scientists came forward, criticizing the integrity of the data environmentalists cited in their pleas for stronger environmental protections. "In science, the experiment must be designed to *test* the hypothesis, not to prove it," writes entomologist William Hazeltine. And this, he argues, is precisely where the "scientist-activists" of the early environmental movement, in their zeal to prove environmental threats, failed. Hazeltine recounts experiments that yielded implausible data yet bypassed meaningful scrutiny because their overall results—evidence that pesticides harm birds—proved politically expedient. To underscore his conviction that scientist-activists have subordinated scientific standards to an environmental agenda, he quotes community organizer Saul Alinsky: "If the end does not justify the means, what the hell does?"

Environmentalists deny that the data upon which they based their predictions are unsound. They argue that on the question of whether or not an environmental crisis threatens humanity, the scientific community is characterized by solidarity; scientists overwhelmingly express the conviction that catastrophe is imminent. "One should not make the mistake of thinking that this is a scholarly controversy," writes scientist Paul Ehrlich of Stanford University. The *World Scientists' Warning to Humanity*, he notes, which warns that the planet's resources are being pressed to their limits—as demonstrated by such humanmade phenomena as increasing greenhouse emissions, ozone depletion, acid rain, declining biodiversity, and deforestation—was "signed by over 1,600 of Earth's leading scientists, including more than half of all living Nobel laureates in science." Ehrlich dismisses those who disagree with this prognosis as "the few scientist contrarians and the gaggle of journalists."

Scientific analyses of the planet's health continue to attract debate and a variety of interpretations among politicians, environmental activists and their opponents, and members of the scientific community. This chapter features authors who have examined the evidence and reached broadly divergent conclusions about whether there is an environmental crisis.

"Our worldwide civilization confronts an unprecedented global environmental crisis."

Environmental Conditions Are Deteriorating

Al Gore

Al Gore has served as a senator from Tennessee and as vice president of the United States and is the author of *Earth in the Balance: Ecology and the Human Spirit*. The following viewpoint is adapted from a speech Gore delivered to the United Nations Commission for Sustainable Development at its first session. Gore argues that there is a serious need for immediate environmental conservation and preservation efforts. Overpopulation, pollution, and other problems affecting the environment are reaching crisis levels worldwide, he contends. If the world community fails to take measures to control these problems, Gore predicts, environmental disaster is inevitable.

As you read, consider the following questions:

1. According to Gore, how does overpopulation adversely affect the environment?
2. In the author's opinion, what factors threaten the world's freshwater supply?
3. What two principles does Gore believe nations should follow in order to prevent environmental disaster?

From a reprint of excerpts of Al Gore's speech to the first session of the United Nations Commission for Sustainable Development, in *Our Planet*, vol. 6, no. 2, 1994.

In 1992 at the United Nations Conference on the Environment (UNCED) in Rio de Janeiro, the great riches of human creativity were on full display. Scientists displayed startlingly beautiful computer images of every square inch of the Earth—as seen from space. Artists crafted spectacular sculptures, paintings, music, graphics and films. And they all seemed more alike than different: the indigenous person and the artist, the scientist and the child, the tourist and the diplomat. All seemed to share a deeper understanding—a recognition that we are all part of something much larger than ourselves, a family related only distantly by blood but intimately by commitment to each other's common future.

And so it is today. We are from different parts of the globe. But we are united by a common premise: that human activities are needlessly causing grave and perhaps irreparable damage to the global environment. And the dangers are clear to all of us.

Degradation of land, forests and fresh water—individually and synergistically—plays critical roles in international instability. Huge quantities of carbon dioxide, methane, and other greenhouse gases dumped in the atmosphere trap heat, and raise global temperatures. Harvard Professor Edward Wilson, writing in the *New York Times*, summarized the notions of those who have a different view.

'Population growth? Good for the economy—so let it run. Land shortages? Try fusion energy to power the desalting of sea water, then reclaim the world's deserts . . . by towing icebergs to coastal pipelines. . . .

'Species going extinct'? Not to worry,' the skeptics say. 'That is nature's way. Think of humankind as only the latest in a long line of exterminating agents in geological time. Resources? The planet has more than enough resources to last indefinitely.'

Wilson called this group the 'exemptionalists' because they hold that humans are so transcendent in intelligence and spirit that they have been exempted 'from the iron laws of ecology that bind all other species.'

No Exemption for Humans

The human race is not exempt. The laws of ecology bind us, too. We made a commitment at Rio to change our course. We made a commitment to reject the counsel of those who would continue along the road to extermination.

But of course, what we have done so far is only a beginning. We cannot overestimate the difficulties that lie ahead. In fact, from the vast array of problems about which it is possible to be pessimistic, let me mention two.

First, population growth. It is sobering to realize what is happening to the world's population in the course of our lifetimes.

From the beginning of the human species until the end of World War II, when I was born, it took more than 10,000 generations to reach a world population of a little more than 2 billion. But in just the past 45 years, it has gone from a little over 2 billion to 5.5 billion. And if I live another 45 years, it will be 9 or 10 billion.

Today's Problems

The changes brought about by this explosion are not for the distant future. This is not only a problem for our grandchildren. The problems are already here. Soil erosion. The loss of vegetative cover. Extinction. Desertification. Famine. The garbage crisis.

The population explosion, accompanied by wholesale changes in technology, affects every aspect of our lives, in every part of the globe.

Now, sometimes, developing countries feel the population argument is one made by wealthy countries who want to clamp down on their ability to grow.

Only a Child

I'm only a child, but I'm very worried—because I hear about the thinning ozone layer, global warming, forests disappearing and the oceans being polluted. I can't control these things but they will hurt my life. . . .

Perhaps, as a child, my life and ideas are pretty simple and straightforward, but I wonder sometimes if adults, in their complicated work and lives, forget what's really important.

Is the secret for adults to find real values, to remember all the insects and birds of their own childhood, to remember catching butterflies and looking for frogs in ponds? To remember playing in the grass and climbing trees? To remember how important these things were, how the world would be unimaginable without them? And to remember how they trusted grown-ups to make sure everything would be all right?

Severn Suzuki, *Our Planet*, vol. 6, no. 2, 1994.

Sometimes the developing countries are right. So I say this to citizens of the developed nations: we have a disproportionate impact on the global environment. We have less than a quarter of the world's population—but we use three quarters of the world's raw materials and create three quarters of all solid waste. One way to put it is this: a child born in the United States will have 30 times more impact on the Earth's environment during his or her lifetime than a child born in India.

The affluent of the world have a responsibility to deal with their disproportionate impact. But population growth affects everyone. By the year 2000, 31 low-income countries will be unable to feed their people using their own land.

Is population growth only a problem of birth control? Of course not. Paradoxically, reducing infant mortality is important as well. Several decades ago, Julius Nyerere [Tanzania's first elected president] put this matter cogently: 'The most powerful contraceptive is the confidence of parents that their children will survive.'

Slowing population growth is in the deepest self-interest of all governments. It is a responsibility for rich and poor countries alike.

Rapid population growth is only one of the causes of a profound transformation in the relationship between human civilization and the ecological system of the Earth. The emergence of extremely powerful new technologies that magnify the impact each of us can have on the global environment has also played an important role.

Most significant of all, many people now think about our relationship to the Earth in ways that assume we don't have to concern ourselves with the consequences of our actions—as if the global environment will forever be impervious to the rapidly mounting insults to its integrity and balance. But the evidence of deterioration is all around us.

Take, for example, the threat to our supply of fresh water.

There is a lot of water on Earth. But there isn't very much fresh water—only about 2.5 percent of all water on Earth is fresh and most of that is locked away as ice in Antarctica or Greenland, or other areas.

Furthermore, much of that water is used inefficiently. It may also be polluted by toxics and human waste. Meanwhile, by the year 2000, 18 of the 22 largest metropolitan areas in the world—those with more than 10 million people—will be in developing countries. By 2025, 60 percent of the world's population will live in cities—that's more than 5 billion people. They will urgently need fresh water and water sanitation.

Even though our worldwide civilization confronts an unprecedented global environmental crisis, we go from day to day without confronting the rapid change now under way. We must recognize the extent to which we are damaging the global environment, and we must develop new ways to work together to foster economic progress without environmental destruction.

> "There is every scientific reason to be joyful
> about the trends in the condition of the Earth,
> and hopeful for humanity's future."

Environmental Conditions Are Improving

Julian L. Simon

In the following viewpoint, Julian L. Simon contends that there is no environmental crisis, nor is there likely to ever be one. Simon expresses his conviction that, far from suffering degradation, environmental conditions are in fact improving and show every indication of continuing to do so into the foreseeable future. Pessimistic predictions of environmental conditions are not supported by accurate scientific research, he concludes. Simon is a professor of business administration at the University of Maryland and the author of *The Ultimate Resource* and *The Resourceful Earth: A Response to the Global 2000 Report*.

As you read, consider the following questions:

1. What does the author cite as indicators of a healthy environment?
2. Which scientific research does Simon cite to support his ideas? What specific data does he relate?
3. What were the findings of a 1986 National Academy of Sciences report on world population, according to Simon?

Julian L. Simon, "Doing Fine on Planet Earth," *Washington Times*, A-23, April 21, 1995. Reprinted by permission of the author.

Tomorrow [April 22, 1995] marks the 25th anniversary of Earth Day. Now as then, its message is spiritually uplifting. But all reasonable persons who look at the latest statistical evidence must agree that Earth Day's scientific premises are entirely wrong.

Environmentalists' False Alarm

During the first great Earth Week in 1970 there was panic. The doomsaying environmentalists—of whom the dominant figure was Paul Ehrlich—had raised the alarm: The oceans and the Great Lakes were dying; impending great famines would be seen on television starting in 1975; the death rate would quickly increase due to pollution; and rising prices of increasingly scarce raw materials would lead to a reversal in the past centuries' progress in the standard of living.

The media trumpeted the bad news in headlines and front-page stories. Mr. Ehrlich was on the Johnny Carson show for an unprecedented full hour—twice. Classes were given by television to tens of thousands of university students.

It is hard for those who did not experience it to imagine the national excitement then. Even those who never read a newspaper joined in efforts to clean up streams, and the most unrepentant slobs refrained from littering for a few weeks.

Population growth was the great bugaboo. Every ill was the result of too many people in the United States and abroad. The remedy, doomsayers urged, was government-coerced birth control, abroad and even at home.

On the evening before Earth Day, I spoke on a panel at the jam-packed auditorium at the University of Illinois. The organizers had invited me to provide "balance." I spoke then exactly the same ideas that I write today; some of the very words are the same.

Of the 2,000 persons in attendance, probably fewer than a dozen concluded that anything I said made sense. A panelist denounced me as a religious nut, attributing to me weird beliefs, such as that murder was the equivalent of celibacy. My 10-minute talk so enraged people that it led to a physical brawl with another professor.

Every statement I made in 1970 about the trends in resource scarcity and environmental cleanliness turned out to be correct. Every prediction has been validated by events. Yet the World Bank, the United Nations Fund for Population Activities, the U.S. Agency for International Development (AID), the CIA and the Clinton administration still take as doctrine exactly the same ideas expressed by the doomsayers in 1970.

Here are the facts: On average, people throughout the world have been living longer and eating better than ever before. Fewer people die from famine nowadays than in earlier centuries. The real prices of food and of every other raw material are lower now than in earlier decades and centuries, indicating a trend of increased natural-resource availability rather than increased scarcity. Major air and water pollutants in the advanced countries have been lessening rather than worsening.

In short, every single measure of material and environmental welfare in the United States has improved rather than deteriorated. This is also true of the world taken as a whole. All the long-run trends point in exactly the opposite direction from the projections of the doomsayers.

Rising Standards of Living

The premise that our first priority should be to do more for our descendants is debatable. Surely it is ethically relevant that our grandchildren will in all likelihood be much better off than we are. While nobody can accurately predict long-term growth rates, remember that standards of living are three times higher than 60 years ago in the United States, seven times higher in Germany and almost ten times higher in Japan. Should my American grandparents have reduced their standard of living, when life was considerably more nasty, brutish and short than now, to leave raw materials in the ground for my benefit?

Lawrence H. Summers, *The Economist*, May 30, 1992.

There have been, and always will be, temporary and local exceptions to these broad trends. But astonishing as it may seem, there are no data showing that conditions are deteriorating. Rather, all indicators show that the quality of human life has been getting better.

As a result of this evidence of improvement rather than degradation, in the past few years there has been a major shift in scientific opinion away from the views the doomsayers espouse. There now are dozens of books in print and hundreds of articles in the technical and popular literature reporting these facts.

Responding to the accumulating literature that shows no negative correlation between population growth and economic development, in 1986 the National Academy of Sciences (NAS) published a report on population growth and economic development prepared by a prestigious scholarly group. It reversed almost completely the frightening conclusions of the previous 1971 NAS report. The group found no quantitative statistical evidence of population growth's hindering economic progress, though they hedged their qualitative judgment a bit. The report found benefits of additional people as well as costs. Even the World

Bank, the greatest institutional worrier about population growth, reported in 1984 that the world's natural resource situation provides no reason to limit population growth.

Betting on a Better Environment

A bet between Mr. Ehrlich and me epitomizes the matter. In 1980, 10 years after the first Earth Day, Mr. Ehrlich and two associates wagered with me about future prices of raw materials. We would assess the trend in $1,000 worth of copper, chrome, nickel, tin and tungsten for 10 years. I would win if resources grew more abundant and thus cheaper, and they would win if resources became scarcer and thus more expensive. At settling time in 1990, the year of the 20th Earth Week, they sent me a check for $576.07.

A single bet proves little, of course. Hence I have offered to repeat the wager, and I have broadened it as follows: I'll bet a week's or a month's pay that any trend pertaining to material human welfare will improve rather than get worse. You pick the trend—perhaps life expectancy, the price of a natural resource, some measure of air or water pollution, or the number of telephones per person—and you choose the area of the world and the future year the comparison is to be made. If I win, my winnings go to nonprofit research.

I have not been able to close another deal with a prominent academic doomsayer. They all continue to warn of impending deterioration, but they refuse to follow Professor Ehrlich in putting their money where their mouths are. Therefore let's try the chief "official" doomsayer, Vice President Al Gore. He wrote a best-selling book, *Earth in the Balance*, that warns about the supposed environmental and resource "crisis." In my judgment, the book is as ignorant and wrongheaded a collection of cliches as anything ever published on the subject.

So how about it, Mr. Vice President? Will you accept my offer? And how about your boss, Bill Clinton, who supports your environmental initiatives? Can you bring him in for a piece of the action?

It is not pleasant to talk rudely like this. But a challenge wager is the last refuge of the frustrated. And it is very frustrating that after 25 years of the anti-pessimists being proven entirely right, and the doomsayers being proven entirely wrong, the latter's credibility and influence wax ever greater.

That's the bad news. The good news is that there is every scientific reason to be joyful about the trends in the condition of the Earth, and hopeful for humanity's future, even if we are falsely told the outlook is grim. So Happy Earth Day.

"*Many oft-cited 'facts' used to paint a picture of impending ecological disaster are more myth than reality.*"

The Environmental Crisis Is Exaggerated

Stephen Budiansky

Environmentalists have raised concern about the prospect of global warming and holes developing in the ozone layer. In the following viewpoint, Stephen Budiansky contends that these "scientist-activists" have falsely alarmed the public, warning of impending crises for which definitive evidence is yet to be discovered. In doing so, Budiansky argues, environmental groups have damaged their credibility. Stephen Budiansky is a senior writer for *U.S. News & World Report* weekly magazine.

As you read, consider the following questions:

1. What must environmental scientists do, according to Budiansky, if they are to restore their credibility?
2. If data collected by NASA researchers regarding ozone levels above Maine are accurate, as Budiansky asserts, why does he consider the predictions based on this information flawed?
3. What two major problems does the author find with the current computer models used to assess global warming trends?

From "The Doomsday Myths" by Stephen Budiansky, *U.S. News & World Report*, December 13, 1993; ©1993, U.S. News & World Report. Reprinted with permission.

It used to be that only religious cranks could predict the end of the world with a straight face. No more. Eminent scientific forecasts of imminent global ecodeath have become a routine part of the debate over environmental protection. A petition signed by 1,575 leading scientists says that man's activities may render Earth "unable to sustain life." Vice President Al Gore, in announcing the Clinton administration's environmental initiatives, regularly invokes an equally apocalyptic vision of the future: mass extinctions, cataclysmic climate upheavals, gaping ozone holes. "We have no right to cavalierly risk the ecological balance of this Earth," Gore said recently in an interview with *U.S. News.*

Losing Credibility

But such warnings of impending doom have come under furious counterattack. Late 1993 brought a spate of books and articles, most written by conservative academics and columnists, that dismiss all warnings of environmental doom as hoaxes. Columnist George Will has derided Gore's concern for the planet as a government power grab; conservative talk-show host Rush Limbaugh has called global warming a "scam" invented by environmental scientists to increase their research funding, and books with titles like *Apocalypse Not, Trashing the Planet* and *Environmental Overkill* have joined the fray, purporting to debunk as pseudoscience and "scaremongering" everything from acid rain to air pollution and toxic waste.

Coming at a time when Americans are preoccupied with foreign competition, jobs and the deficit, the backlash is having an effect. Although Gore characterizes skepticism on global warming as "unethical" and based upon "kooky theories," industry opposition to tough global-warming measures played a key role in the administration's compromise plan, which relies almost entirely on voluntary action. Industry has succeeded in getting the administration to compromise on other important environmental issues: The White House in 1993 dropped plans to increase fuel-economy standards for automobiles and allowed logging to continue in 25 percent of old-growth forests—decisions that have outraged environmentalists.

What's driving the backlash? Certainly, none of the global environmental issues now under attack is a hoax. Nor is the political agenda of many of the anti-environmentalists very hard to find. But some environmental researchers now concede that at least part of the blame lies with themselves: Environmentalists' penchant for doomsaying is coming back to haunt them. By overstating evidence, by presenting hypotheses as certainties and predictions as facts to create a sense of urgency, scientist-activists have jeopardized their own credibility. "The front lash," acknowledges Stephen Schneider, a leading climate modeler with the National Center for Atmospheric Research in Colorado, "was a little political, too."

Others, while denying that environmental scientists have deliberately played politics with their data, concede that the distinction between science and advocacy has become blurred as scientists are increasingly called on to make policy recommendations—which often means playing down scientific uncertainties. "It's one thing to discuss these things with scientific colleagues; it's another to sit before a panel of policy makers and say this is what we ought to do," says Tom Lovejoy of the Smithsonian Institution, who has advocated urgent action to protect tropical biodiversity. "When you live in a sound-bite world, they don't stand still if you wobble around."

A review of the scientific literature and interviews with researchers suggest that while none of the threats to the global environment can be dismissed, many oft-cited "facts" used to paint a picture of impending ecological disaster are more myth than reality. Only by confronting these myths, some researchers now say, can environmental scientists hope to retain their credibility in the face of mounting skepticism—and get on with addressing the real environmental challenges the world faces. Keeping the water clean, in other words, is the best way to make sure the baby doesn't get thrown out with it. . . .

The Myth of Ozone Depletion

After two decades of probing the atmosphere with a welter of instruments carried on spacecraft, weather balloons and highflying U-2 aircraft, researchers thought the case was closed: Manmade chemicals known as chlorofluorocarbons (CFCs) drift into the stratosphere, break down and release chlorine molecules, which in turn attack the ozone molecules that naturally shield Earth from cancer-causing ultraviolet radiation.

Now, a furious counterattack has caught many researchers by surprise. Former Atomic Energy Commission Chairman Dixy Lee Ray, in her book *Trashing the Planet*, claims that any ozone depletion that has occurred is completely natural, the result of volcanic eruptions and sea spray, not man-made chemicals.

Most of these counterarguments are easily dismissed. Measurements in the atmosphere confirm what basic chemistry suggested all along: Chlorine from most natural sources never reaches the stratosphere. Natural chlorine-bearing chemicals, such as hydrochloric acid from volcanoes and sodium chloride from sea spray, are soluble in water and are washed out of the air by rain. Measurements taken after the 1991 Mount Pinatubo volcanic eruption in the Philippines showed no increase of atmospheric chlorine.

The exception is when volcanoes erupt with such force that they inject material directly into the stratosphere, which begins at an altitude of about 60,000 feet. The 1976 eruption of Mount St. Augustine was one such exception; it deposited 175,000 tons of chlorine, but that is still much less than the 750,000 tons released each year from man-made chemicals.

Ed Gamble/*Florida Times-Union*. Reprinted with permission.

The only natural source of chlorine that routinely survives the journey to the stratosphere is methyl chloride, which is given off by ocean plankton; measurements show it accounts for only one sixth of the chlorine in the stratosphere. All the rest comes from man-made CFCs, and stratospheric chlorine has been increasing steadily since the 1950s, from 0.6 parts per billion (ppb) to 3.8 ppb now.

But if Ray and others have exaggerated how much chlorine enters the stratosphere from natural sources, environmental activists have overstated the proven consequences of man-made CFCs. A number of ill-substantiated claims of increases of skin cancer and epidemics of blindness in sheep in South America have received much attention. So did a January 1992 announcement by National Aeronautics and Space Administration (NASA) researchers of a record high measurement of ozone-eating chlorine molecules made by a U-2 aircraft over Bangor, Maine. The data were correct—but subsequent interpretations of them were not. Then Senator Gore chided George Bush for ignoring an "ozone hole over Kennebunkport," where the former president has a vacation home. The findings were also cited by advocates urging a speedup of the already agreed-to international phaseout of CFC production.

But the predicted ozone hole in the Northern Hemisphere never materialized, and the NASA researchers, fairly or unfairly, ended up with egg on their faces. The reason was a warming trend in late January 1992. Polar winter temperatures usually accelerate ozone loss: Nitric acid, which in warmer conditions forms an atmospheric gas that binds to some of the chlorine to keep it out of circulation, condenses at -78 degrees Celsius, and chlorine concentrations then shoot up. The January 1992 thaw reversed the process, and chlorine levels dropped.

"Poor Science"

"It's very poor science to assume that ozone is dropping based on circumstantial evidence of increased chlorine," acknowledges Harvard's Jim Anderson, one of the NASA-funded researchers. "People who say we have not established a cause-and-effect link in the Northern Hemisphere are correct." In the Antarctic, measurements taken by Anderson and others have confirmed the connection between elevated chlorine and depleted ozone. Antarctic ozone holes have appeared every spring since the mid-1970s and have been growing deeper each year; up to 80 percent of the ozone disappears over an area equal to 10 percent of the globe's surface.

But the Antarctic ozone hole is the product of *two* factors: man-made chlorine and extreme cold. Although man-made chlorine is distributed throughout the stratosphere, extreme cold is confined to polar winters. So the existence of the Antarctic hole does not in itself prove that severe ozone depletion also will occur in more temperate regions.

In fact, some natural processes may counteract the ozone-depleting effects of CFCs at midlatitudes, especially in the Northern Hemisphere—where most of the world's population is concentrated. The large number of mountain ranges in the Northern Hemisphere acts to stir up the atmosphere, pumping heat to the North Pole; the Arctic stratosphere stays about 10 degrees warmer than the Antarctic. That prevents chlorine-binding nitric acid from condensing out of the air. Air currents also work to shuttle ozone from the tropics to the midlatitudes, replacing some of the loss.

The fact that the stratosphere is higher over the midlatitudes than it is over the poles poses a fundamental problem for researchers trying to make the case for man-made ozone depletion over the temperate zones. Because the stratosphere there lies beyond the reach of U-2 aircraft, direct, simultaneous measurements of ozone and chlorine are difficult. (NASA has proposed buying an unmanned, highflying aircraft to gather the needed data.)

Much depends on the answer. The industrialized nations have agreed to terminate CFC production by 1996, but China, India and Brazil have been given an additional 10 years to comply, and all three will be potentially huge producers as they industrialize over the next decade. It may turn out that the stratosphere can absorb the blow and the damage will remain confined to the Antarctic in winter. But if not, the damage will persist for a very long time: The stratosphere flushes itself out very slowly, and the chlorine already there will persist for hundreds of years.

The Myth of Global Warming

To environmentalists, global warming is "a holocaust" or "the end of nature." To the political right, it is a trojan horse for expanded government powers.

The issue started to heat up politically in the 1980s, when Senator Gore began to argue that computer models forecasting a warming of the planet were not merely theoretical predictions; the scientific community, he said, accepted their results as near certainty. Those who still question this, Gore recently told *U.S. News*, are suffering "a massive case of denial" and have "willfully put out false, scientific pseudo facts to pollute the public debate." He added: "Some of the so-called scientists who put out these kooky theories get money from industries that profit greatly from the current pattern . . . there are institutes funded by coal companies that have an interest in not seeing any change take place."

Yet doubts about the likelihood, intensity and consequences of global warming extend far beyond a few fringe scientists or industry hirelings.

This much is certain: Carbon dioxide, water vapor and several other atmospheric gases trap heat that otherwise would radiate from Earth, leaving the planet to freeze. And since 1750, the concentration of carbon dioxide in the atmosphere has increased from 275 ppm to 355 ppm as the burning of fossil fuels has expanded. (The data come from direct atmospheric measurements since 1958; air trapped in the ice that builds up each year in polar ice sheets has preserved a record of earlier years.)

Beyond such basic facts, however, uncertainty reigns. Computer models of the atmosphere calculate that a doubling of the carbon dioxide concentration, which could occur in the next century if no action is taken to limit emissions, will cause a 1-to-5-degree Celsius rise in average global temperatures.

But there are two major problems with the models. First, the current models account only very roughly for the large role the oceans play in determining the flow of heat in the atmosphere. Second, as physicist Richard Lindzen of the Massachusetts Institute of Technology, a persistent critic, notes, most of the temperature rise calculated by the models comes about not from the increase in carbon dioxide itself but from an increase in the concentration of water vapor in the air, which the models calculate as a secondary consequence of rising temperatures.

Lindzen argues that there is no physical theory or evidence to support the assumption that water vapor increases in the atmosphere as temperature rises. "The only reason to think that is that warm air *can* hold more water vapor," he says. "That's equivalent to saying, 'I have a 2-liter beer mug, you have a 1-liter mug, so mine will always have more beer in it.'". . .

Reality Check

That has led both proponents and critics of the global-warming theory to use historical temperature records to try to make their case. Proponents note that average global temperatures have risen in the past century. Skeptics counter that the temperature increase, about 0.5 degrees Celsius, is less than half what the models predict given the carbon dioxide increase to date.

The problem with all such analyses is that *natural* cycles of warming and cooling are always occurring. Since the end of the Ice Age 15,000 years ago, there have been one or two centuries in each millennium in which temperatures have risen half a degree. That means there is a 10 to 20 percent chance that the temperature rise in the past century is part of a natural warming cycle and has nothing to do with carbon dioxide. On the other hand, says climate modeler Stephen Schneider, "there's an equal chance that we're in a natural cooling trend now, and thus our effect is twice as big."

Trying to spot a general warming trend while it's happening is almost impossible, says Harvard planetary scientist Michael McElroy: "I really do not believe we're likely to suddenly have someone say, 'Eureka, the greenhouse is here!' If we ever do get to that point, we'd already be in bad shape."

But even some of the strongest skeptics agree that it is sensible to reduce carbon dioxide emissions for reasons that have nothing to do with the greenhouse effect. Fred Singer, a professor of atmospheric chemistry at the University of Virginia and a vocal critic of global-warming predictions, argues that an increase in the gasoline tax, energy conservation and improvements in energy efficiency all make sense under any circumstances. That kind of common sense is much less dramatic than predicting "the end of nature." But it may serve both science and the environment better in the long run.

> *"Scientists who have sown seeds of doubt about global warming . . . 'are not among the more capable scientists in the field.'"*

The Environmental Crisis Is Real

Barbara Ruben

An "environmental backlash" is taking place, *Environmental Action* magazine editor Barbara Ruben argues in the following viewpoint. According to Ruben, reputable news media have joined the ranks of those inaccurately reporting that environmental problems do not exist. Ruben criticizes these journalists for failing to fully examine the source of the data they cite when making such claims. Instead, she maintains, reporters should expose the relationship between the research organizations that generate inaccurate information and the industries that fund them and stand to gain from the findings produced.

As you read, consider the following questions:

1. What important information does Ruben accuse some reporters of omitting?
2. How has public relations affected news coverage of the environment, according to Jim Naureckas, as quoted by the author?
3. According to Ruben, what scientific information does Dixie Lee Ray overlook when claiming that CFCs pose no threat? How does F. Sherwood Rowland, as quoted by Ruben, refer to Ray's research in this instance?

Excerpted from "Back Talk" by Barbara Ruben, *Environmental Action*, Winter 1994. Reprinted with permission.

Exposure to the chemical dioxin may be no more dangerous than spending a week sunbathing. Global warming is benign. Taxpayers spend billions battling environmental problems that pose little danger. What ozone hole?

All these assertions by reporters made their way into news articles in the venerable *New York Times* and *Washington Post* in 1993 and 1994. Bolstered by quotes and figures by legions of scientists and policymakers, the news reports sound like official pronouncements that the planet has never been better.

An Environmental "Backlash"

And more environmental "backlash" books have been published in 1993 than ever before. With names like *Eco-Scam* and *Environmental Overkill*, the books are written by an array of journalists, economists and scientists and purport to "debunk" the gamut of environmental issues, from acid rain to ozone depletion.

Although these books more overtly ridicule environmentalists as hysterics bent on hyping far-fetched apocalyptic theories than do newspaper reports, both hammer home the same message: environmentalism has gone too far.

Writes the *New York Times'* Keith Schneider in the first sentence of a 5-part, 12,000-word series the newspaper prominently featured in March 1994, "many scientists, economists and government officials have reached the dismaying conclusion that much of America's environmental program has gone seriously awry."

But what really may be awry is the skewed way in which the books and some news articles twist science to their advantage, focusing on a study or source that flies in the face of established scientific research or vaguely attributing hypotheses to "some scientists say" and "experts agree." Few of the reporters and authors question the motivations of the scientists whose research is quoted, rarely attributing a study's funding source or institution's political slant. . . .

Balanced News or PR?

Says Schneider of his reporting style, "The kind of questions I've begun to ask have been prompted by my reporting in the field, where people are asking whether in an era of financial limits we're doing enough to understand these risks and to invest the money properly and wisely and get our money's worth out of environmental protection."

He also told the *American Journalism Review* in the summer of 1994, "What drives the national environmental groups is not necessarily the truth. Environmental journalists have to regard environmental groups with as much skepticism as we have traditionally regarded polluters. . . . We haven't done enough to look at the other side [of the issues]. We haven't done as good a job to find those scientists who are skeptical.". . .

Jim Naureckas, editor of *EXTRA!*, the bimonthly magazine of the media watchdog group Fairness and Accuracy in Reporting, says that in the rush to "balance" a story journalists sometimes grope for an opposing point of view, no matter how little credence it may get from the larger scientific community.

"PR has invaded science. Now we have the proliferation of groups with innocuous sounding names like the U.S. Council on Energy Awareness [funded by the nuclear industry] and the American Council on Science and Health [which has championed the view that PCBs, Alar and asbestos are all harmless in low doses and counts dozens of industry giants such as Dow, Monsanto and Exxon among its contributors]. Journalists don't often distinguish between research put forth by these groups and independent research scientists.". . .

Differing Reports on Ozone

April 15, 1993, was a tumultuous day for ozone. Readers in Dallas and Washington, D.C., opened their newspapers to see such headlines as "Ozone Problem Seems to Be Headed for a Solution" and "After 2000, Outlook for the Ozone Layer Looks Good." But Los Angeles residents read the opposite headline, "Wider Damage to the Earth's Ozone Layer Is Feared," and those in New York learned that "Satellite Finds Growing Threat to Ozone."

The more positive headlines came from a *Washington Post* news service story by Boyce Rensberger. In the *Post*, the story took top billing and continued for a full page and a half inside the paper. It began, "After nearly a decade of headlines and hand-wringing about erosion of the Earth's protective ozone layer, the problem appears to be well on the way to solution." Rensberger went on to optimistically write that "researchers say the problem [ozone depletion] appears to be heading toward solution before they can find any solid evidence that serious harm was or is being done." Because of the Montreal Protocol—the international treaty that bans CFCs—the world has averted the "dark scenarios of environmental doom that were pronounced during the discovery of the Antarctic ozone hole," Rensberger writes.

While the *Washington Post* sunnily projected a virtually risk-free future for the ozone layer, the *Los Angeles Times* painted a much bleaker picture, reporting that "ozone loss over the poles could already be depleting the protective form of oxygen in the upper atmosphere over a much wider area than previously believed."

Scientists Grow Alarmed

So alarmed was the normally unflappable American Chemical Society by Rensberger's article—and characterizations by Rush

Limbaugh, Dixie Lee Ray and other anti-environmental authors that ozone depletion posed no threat—it published an 11-page report in its weekly magazine, *Chemical and Engineering News*. The magazine has a circulation of 130,000 chemists and chemical engineers—more than half working in industry.

A Growing Threat

All the signs are that ozone depletion is intensifying as the 1990s progress. . . .

The new evidence is, indeed, so compelling that in response to the growing threat, the forty-nine original signatories to the Montreal Protocol have now called for a total ban on CFC production by the year 2000. But even if this is achieved, the atmosphere will still contain more CFCs in the year 2000 than it does today, and 90 percent of the CFCs in the air when the ban comes into force will still be there in 2010. The worldwide depletion of stratospheric ozone, not just at the poles but literally over our heads, will get worse before it gets better.

John Gribbin, *The Hole in the Sky*, 1993.

"Ozone researchers are puzzled by the recent tone of news coverage and commentary implying not only that the problem of ozone depletion is solved but that there may never have been a problem," wrote reporter Pamela Zurer in the May 23, 1993, edition.

In an interview, Zurer said that she and her editors decided to write the article after getting calls and letters from readers who had seen conflicting reports in the media. "We got questions from our readers that you wouldn't expect from someone with an advanced degree in chemistry. And if they're confused, I know the general public must be as well."

Says Rensberger, "I was trying to write up to my readers and not write down to them as we reporters often do. . . . I think there is a tradition in environment writing of giving an unquestioning, alarmist spin to the story."

Misreading Science

One book that has fueled debate on ozone [since its 1992 publication] is *Holes in the Ozone Scare: The Scientific Evidence That the Sky Isn't Falling*. Publicity materials for the book advocate "overthrow of the murderous environmentalist regime now ruling our schools, government institutions and the media." In *Environmental Overkill*, Ray cites the book to prove her point that the evidence for banning CFCs is "flimsy and dangerous." Rush

Limbaugh also draws on the book to blast "environmental wackos" out to ruin the American way of life by banning CFCs.

Scientific claims in the books "often rely on misreadings of old scientific literature," says the American Chemical Society's Zurer. "They ignore more recent research that leads to different conclusions than the ones they promote."

Ray, for example, claims that because CFCs are heavier than air they cannot rise into the stratosphere, the critical level in the atmosphere that houses the ozone layer. But according to F. Sherwood Rowland, one of the scientists who first theorized about the damage CFCs could do to the ozone layer, winds in the atmosphere evenly disperse all molecules in the air, no matter what they weigh. And CFCs have been found in measurable amounts in the stratosphere throughout the late 1980s and the 90s.

Rowland, president of the American Association for the Advancement of Science, was so dismayed by the amount of "ignorance and misinformation" in the media and backlash books that he made science communications the subject of his address at the association's 1993 annual meeting. He faults Ray for relying on "fourth-hand descriptions" rather than going to original scientific literature.

And in a book called *Apocalypse Not*, published by the libertarian think tank the Cato Institute, the authors, professors of economics and chemistry at Rhodes College in Memphis, claim natural phenomena, such as volcanoes, far outweigh any effects on the ozone caused by humans. However, research shows that hydrogen chloride emitted from volcanoes rarely reaches the stratosphere and is instead washed out by rain. After the huge eruption of Mount Pinatubo in 1991, a study by the National Center for Atmospheric Research in Boulder, Colorado, found little change in the amount of hydrogen chloride in the stratosphere. . . .

The Debate over Global Warming

In June 1994, National Geographic's journal, *Research and Exploration*, published a special issue on global warming. One article, titled "Benign Greenhouse" and written by Virginia state climatologist and University of Virginia environmental science professor Patrick J. Michaels, asserted that global warming won't be disastrous and in fact could bring the benefit of longer crop growing seasons.

On June 1 of the same year, Boyce Rensberger at the *Washington Post* picked up the story, right down to the headline, "'Greenhouse Effect' Seems Benign So Far." He attributes this premise to the vague standby "scientists say." The front page article, however, devotes the first 15 paragraphs to Michaels's views before quoting a dissenting opinion by James Hansen, director of the Goddard Institute for Space Studies in New York City.

Hansen, who also wrote an article that appeared in the June issue of *Research and Exploration*, believes that the earth will warm significantly due to increased concentrations of carbon dioxide (CO_2) that have been released by burning fossil fuels. Hansen's views have also been voiced by the 200 scientists on the International Panel on Climate Change.

Science for Hire

Rensberger also failed to mention that Michaels publishes a glossy quarterly magazine, *World Climate Review*. The magazine's entire production costs are underwritten by about $300,000 a year in funds from the Western Fuels Association, a group of coal companies. Burning of coal, of course, is one of the major contributors of CO_2 in the atmosphere. Nor does it mention what Michaels freely admitted when asked in an interview—about 25 percent of his research funding also comes from such industry sources as Edison Electric Institute, the largest utility trade association in the country.

"There's no more room for conflict here than from taking money from EPA [Environmental Protection Agency]," Michaels says. "Everybody has an agenda, let's face it."

But Hansen looks askance at such claims, saying that scientists who have sown seeds of doubt about global warming "generally are not among the more capable scientists in the field. . . . Recent reporting, including that in better newspapers, such as the *New York Times* and the *Washington Post*, leaves the incorrect impression that global warming has become a much more uncertain matter," he says.

The *Post* article pointed out that most of the warming scientists observe is taking place at night. A meeting of atmospheric scientists at the University of Maryland in the summer of 1994 addressed the effects of the rise in minimum temperatures. Although Michaels asserts this rise could be beneficial, atmospheric scientist Dan Lashof of the Natural Resources Defense Council says the consensus at the meeting was that even nighttime warming needs to be mitigated because "ecological destruction is driven as much by night time temperatures as during the day." Lashof says one concern is that as the frost-free zone gets larger as nights grow warmer, more tropical diseases might be seen.

In September 1993, the *New York Times* picked up on the backlash theme with an article titled "Scientists Confront Renewed Backlash on Global Warming." Reporter Wallace K. Stevens quotes *Apocalypse Not*'s assertion that evidence of current or future global warming is "ludicrously small." Not until the 29th paragraph of the 32-paragraph article does Stevens admit that "most climate researchers believe there is a better than even chance that the climate will warm by at least 3.5 degrees over the next century."

A Worsening Environment

Predicting environmental trends in science is a process mired in imprecision and uncertainty. But some scientists and journalists fear that recent reporting and anti-environmental books have elevated that uncertainty to the realm of fact.

"Essentially we're flying blind into the future. Instead of being able to provide us with a reliable radar system, scientists are sitting in the front of the plane staring out the window trying to see obstacles ahead," says the *Boston Globe*'s Diane Dumanowski. She criticizes some reporters for assuming in their stories that it is more risky to take action on environmental issues than to delay. "In my experience uncertainty cuts both ways. Usually I don't see this in the stories we write."

And ultimately the barrage of skewed uncertainties and misinformation could have serious consequences for the environment.

"A little confusion in the public is very important in the political process. If policymakers are getting calls with a lot of contradictory evidence, they're going to say, 'we better wait until we get more proof until we do anything.' A little bit of scientific doubt can be parlayed into paralysis of Congress," says the Environmental Research Foundation's Peter Montague. "And this is very, very important because once the seeds of doubt are sown about whether the planet will ever warm up or if the ozone has holes, these problems are going to be that much worse when we finally get around to dealing with them."

Periodical Bibliography

The following articles have been selected to supplement the diverse views presented in this chapter. Addresses are provided for periodicals not indexed in the *Readers' Guide to Periodical Literature*, the *Alternative Press Index*, or the *Social Sciences Index*.

Gary Benoit — "Warm Earth, Cold Earth," *New American*, May 1, 1995. Available from 770 Westhill Blvd., Appleton, WI 54914.

Lester R. Brown — "Vital Signs," *Buzzworm*, January/February 1993.

Eric Chivian — "The Ultimate Preventive Medicine," *Technology Review*, November/December 1994.

Liane Clorfene-Casten — "The Environmental Link to Breast Cancer," *Ms.*, May/June 1993.

Charlene Crabb — "Soiling the Planet," *Discover*, January 1993.

Murray Feshbach — "Mortal Danger: Health and Environmental Crises in the Former Soviet Union," *American Enterprise*, May/June 1993.

Tee L. Guidotti — "Global Climate Change and Human Ecology," *PSR*, December 1993. Available from BMA House, Tavistock Square, London WC1H 9JR, U K.

Propaganda Review — "Anti-Environmental Propaganda: Special Theme Issue," no. 11, Spring 1994.

Anne Swardson — "A Loss That's Deeper Than the Ocean," *Washington Post National Weekly Edition*, October 24–30, 1994. Available from Reprints, 1150 15th St. NW, Washington, DC 20071.

Lewis Thomas — "Environmental Wake-Up Call," *Final Frontier*, June 1995. Available from 1017 S. Mountain, Monrovia, CA 91016.

John Todd — "Living Machines," *Resurgence*, January/February 1994. Available from 33 E. Minor St., Emmaus, PA 18049.

Larry Wilson — "Environmental Destruction Is Hazardous to Your Health," *Social Policy*, Summer 1994.

How Serious Is Air and Water Pollution?

Chapter Preface

When people hear air pollution mentioned, they tend to think of the air outdoors that everyone breathes. Smokestacks spewing dark, gaseous substances into the sky may come to mind, or perhaps the dirty exhaust exhaled from a weather-beaten truck rambling along the highway. Rarely do people think of the place where, according to some studies, breathing can be most hazardous to their health: indoors.

"Sick building syndrome" is a recently discovered phenomenon, identified as the cause of a constellation of symptoms that can range from headaches and nausea to more serious maladies such as asthma and hypersensitivity pneumonitis. Scant data is available, although several scientists have hypothesized that building ventilation systems, which can function as both conduit and host to numerous pollutants, may be the source of the problem. Air-conditioned structures, according to current research, produce a consistently higher prevalence of sick building syndrome than do those that are naturally ventilated or do not mechanically alter air temperatures.

Retrofitting places of work to improve indoor air quality, even when expensive, is the most cost-effective response to sick building syndrome, according to the Environmental Protection Agency (EPA). The EPA contends that avoiding the cost of remedy means accumulating the growing expense of productivity slippage and workdays lost to illness. The agency estimates that cumulative costs to society attributable to sick building syndrome, in terms of medical expenses and lost wages, reach into "the tens of billions of dollars."

Not everyone agrees that the health risks posed by indoor air pollution merit the costs of complying with existing EPA standards, let alone possible new regulations for sick building syndrome, the existence of which is questioned by some people. "The agency doesn't try to build any perspective on what an acceptable risk is," argues Anthony V. Nero Jr., a senior scientist with the University of California's indoor environment program. He faults the EPA for failing to establish the typical risk level for indoor air. Others, who insist that the expense of meeting regulatory standards is exorbitant, warn of a backlash from the person ultimately footing the bill—the American consumer. The degree of risk posed by polluted air, drinking water, and oceans is debated throughout the following chapter.

"The undeniable reality is that the entire nation
has experienced a dramatic and predictable
improvement in urban ozone air quality."

Air Quality
Is Improving

K.H. Jones and Jonathan Adler

According to K.H. Jones and Jonathan Adler, the authors of the
following viewpoint, air quality in the United States is steadily
improving. They contend, however, that the Environmental Protection Agency uses data that ignores this trend as the basis for
its antipollution regulations. As a result, the authors argue, local
communities are saddled with expensive and ineffective solutions they can ill afford and do not need. Jones heads the Zephyr
Consulting Company, based in Seattle, Washington. Adler serves
as the associate director of environmental studies for the Competitive Enterprise Institute, a Washington, D.C., research organization that advocates private-sector control of the environment.

As you read, consider the following questions:

1. What data do Jones and Adler specifically target as
   "anomalous"?
2. Why is the EPA hesitant to dismiss its existing historical data,
   according to the authors? What measures has the agency
   stated that it needs to take before doing so?
3. How has the Ozone Transport Region (OTR) achieved
   "moderate" classification levels, according to the authors?

Excerpted from "Time to Reopen the Clean Air Act: Clearing Away the Regulatory Smog" by
K.H. Jones and Jonathan Adler, *Cato Policy Analysis*, no. 233, July 11, 1995. Reprinted by
permission of the Cato Institute, Washington, D.C.

United States national policy on the control of urban smog is misguided because it fails to account for current pollution trends and is based on the anomalous meteorological conditions of 1988. Although new data on smog have shown that the trends are continuing downward, the Environmental Protection Agency is doing little to halt regulatory overkill. . . .

Federal Mandates: Comply or Lose Funding

Local officials in selected metropolitan areas around the country are scrambling to meet federal deadlines for compliance with the Clean Air Act Amendments of 1990. For many cities, that meant submitting an acceptable state implementation plan in November 1994, a deadline that was not universally met. Failure to meet federal requirements can result in federal sanctions, including the loss of federal highway funds and the disapproval of future federal construction permits. Not only do those plans have to pass federal muster, but they must satisfy the statutory interpretation of environmental groups as well, since the CAAA include public suit provisions to force the imposition of federal sanctions.

The Environmental Protection Agency, in response to CAAA mandates that the agency essentially wrote, has proposed a wide range of policies, from mandatory carpooling and enhanced inspection and maintenance programs to technology standards for factory emissions and new emissions controls on lawn mowers, snow blowers, chain saws, and the like. Although those policies are costly to businesses and consumers, the EPA claims they are nonetheless reasonable and necessary policies for reducing the threats to public health posed by supposedly poor ozone air quality in America's cities.

The EPA has encountered significant opposition to several of its proposed policies, particularly those that require stringent emissions inspection of motor vehicles. In fact, several states have openly rebelled against implementing those programs within the federally mandated time frame, and Virginia has gone so far as to take the EPA to court.

One area in which the imposition of strong air pollution control measures is particularly contended is the Ozone Transport Region. The OTR consists of 12 states and the District of Columbia, stretching from Virginia up the Atlantic Coast to Maine. The state-level regulatory programs within the OTR are being coordinated through the Ozone Transport Commission, which was established under the CAAA. The OTC has proposed several emissions control strategies for adoption throughout the region. In particular, the OTC has asked the EPA to impose low-emission vehicle standards similar to those imposed in California, which include the use of zero-emission vehicles, otherwise known as electric cars. On December 19, 1994, the EPA approved the OTC request. If a majority of the states in the OTR formally approve of the plan, California-style low-emission vehicle (LEV) standards will become mandatory throughout the entire region. Additional pressure is placed on the states to accept the plan because the EPA has declared that "unless an acceptable LEV-equivalent program is in effect," states in the OTR will be out of compliance with the CAAA, which could possibly result in the imposition of federal sanctions. The OTC is also promoting stringent NO_x control strategies, which are more than likely counterproductive to anticipated O_3 reductions.

Table 1

Total Exceedances at Worst-Case Monitors in California and in Rest of the United States: 1985–94

	1985	1986	1987	1988	1989	1990	1991	1992	1993	1994
California	257	272	270	308	233	167	185	175	128	193[a]
Rest of Country	188	166	275	607	101	124	155	53	104	93[b]
Total	445	438	545	915	334	291	340	228	232	286
Percentage in California	58	62	50	34	70	57	54	77	55	67

[a]Los Angeles maximum = 107 days at worst monitor.

[b]Longest exceedances were 14 days in Houston and 7 days in Dallas.

Policy Analysis No. 223, Cato Institute, July 11, 1995.

Many cities in the OTR are subject to those air pollution control requirements because of the abnormally high air pollution levels that prevailed in 1988, in particular, high levels of urban ozone, or smog. The excess air pollution was caused, in part, by the meteorological conditions that occurred in 1988, by some accounts a once-in-100-years phenomenon akin to the disastrous 1993 floods in the Midwest. For that reason, using data from 1988 in the formulation of air pollution control requirements is misguided, according to a paper by K.H. Jones published by the Cato Institute in 1992. . . .

Air Quality Trends in America

Despite continuing overheated claims by government officials and environmental groups about the threat posed by urban ozone, the undeniable reality is that the entire nation has experienced a dramatic and predictable improvement in urban ozone air quality. Indeed, it is difficult to identify a significant air pollution problem outside California. A broad indication of the improvement is shown in Table 1.

"Exceedances" refers to the number of times that the most highly impacted ozone air quality monitor in a metropolitan area registers ozone concentrations greater than 0.124 parts per million for one hour or more.

California, in particular the Los Angeles Basin, is easily the most heavily polluted area in the United States. In 1994 the worst-case monitor in Los Angeles had 107 exceedances, 14 more than the total number of exceedances (93) for all urban areas in the rest of the nation outside California. The greatest number of exceedances in other cities in 1994 was 14 days in Houston and 7 days in Dallas. The net reduction outside California between 1985–87 and 1992–94 was 57 percent. In California the improvement was 27 percent.

Even a cursory look at the data in Table 1 reveals that the high number of ozone exceedances during 1988 was an anomaly and not an accurate indication of air quality in America's cities. Nonetheless, metropolitan areas' current attainment or nonattainment [in meeting federal air quality standards] status is determined by examining a three-year data window—1988–90—that is frozen in place and does not change to reflect current air quality information. . . .

Most of the nonattainment regions outside California easily meet federal ozone air quality standards today. Those that do not are on the cusp of attainment.

Skewed EPA Standards

The EPA and the OTC argue that it is legitimate to base long-range smog reduction policy on 1988 ozone levels. The EPA has claimed that attempting to account for the meteorological conditions that caused the high number of exceedances in 1988 would be "superficial . . . premature and imprudent." The OTC has maintained that to exclude 1988 in analyzing air quality trends would result in a "misleading" projection. The rationale for accepting 1988 data for today's regulatory purposes is that the ozone exceedances observed that year—almost entirely a function of hot, dry weather—are by no means particularly anomalous and represent just the kind of "outer-bound" event that federal ozone rules are expected to mitigate. As an internal EPA memorandum circulated to counter the 1992 Jones study declared, "There needs to be a certain note of caution. We have had predictions of success before that did not pan out."

To address that concern, we have conducted further analyses

by examining 27 years of detailed day-by-day summertime temperatures for Philadelphia, an area particularly well suited to represent overall trends in the OTR. In addition to daily temperatures and calm winds, the length of high-temperature episodes is an important factor in smog formation in Philadelphia and elsewhere because high temperatures and corresponding ozone levels in the OTR are associated with multiday stagnations caused by Bermuda highs off the East Coast.

Table 2 summarizes the Philadelphia weather data. Examination of those data clearly indicates the unique meteorological circumstances that made 1988 anomalous for policy considerations. The two 18- and 22-day episodes were almost continuous in July and August of 1988. In addition, the anomalous number of 95° and 100° Fahrenheit daytime high temperatures sets 1988 apart. As already noted, attempting to prevent one or more exceedances, equivalent to those of 1988, in a future year is like requiring that levees along the Mississippi be built high enough to prevent all possible future floods.

Table 2

Temperature Statistics and Episode Persistence Profiles for Philadelphia: 1966–93

| Year | Annual Days Above | | | Total Days of Episodes > 3 Days | Number of Episodes > 10 Days | Episode Persistence (days) |
	90°F	95°F	100°F			
1966	33	9	3	32	1	12
1973	28	8	0	25	1	10
1977	26	5	1	22	1	10
1979	17	0	0	14	1	11
1980	36	10	0	31	2	11, 12
1983	45	9	0	41	2	10, 16
1986	33	3	0	28	1	13
1988	52	18	5	49	2	18, 22
1991	53	17	1	51	1	12
1993	39	ND	3	ND	1	11

Notes: Only those years that experienced at least one episode longer than 10 days in length are shown. ND = no data.

Policy Analysis No. 223, Cato Institute, July 11, 1995.

The EPA has, however, acknowledged that weather can have a significant impact on ozone formation in any given year, and the agency argued that "there needs to be a careful analysis to ensure that the high levels associated with the meteorological conditions of 1983 and 1988 are past history and not lurking around the corner." Interestingly, to the authors' knowledge, the EPA has yet to conduct any analysis of historical meteorological data to prove that their base-line air quality level (1988–90) is consistent with the intent of the ambient ozone standard. Although the EPA was required by the Clean Air Act to review the form of the standard, nothing has been done that would compensate for the current reliance on 1988 in the classification of ozone nonattainment areas.

Ozone Air Quality in the Northeast

The OTR was the area outside California most negatively affected by smog in 1988. There has been significant improvement over the past few years in all but a few of the metropolitan areas of the OTR.

The final thing to note is that, given the latest available data, there are no areas in the OTR currently classifiable as "severe" or "serious" for regulatory purposes. Unfortunately, the EPA has seen to it that the 1988-based classifications are cast in stone. One might ask, however, why we should be instituting future control measures that are required to get from a "serious" classification level to attainment when we have already dropped to a "moderate" attainment classification by doing nothing.

"In the United States, the limit of exposure [to unsafe ozone levels] is often exceeded, and it has been estimated that some 75 million people are exposed."

Air Pollution Remains a Major Problem

United Nations Environment Programme

The United Nations established the United Nations Environment Programme (UNEP) to function as a watchdog for world environmental conditions, alerting governments around the globe of impending ecological dangers so that these may be avoided through preventive measures. In the viewpoint below, UNEP argues for greater international cooperation and involvement in research and other efforts to reduce air pollution, which the organization identifies as a "major" global problem. Pointing out that air pollution does not respect national boundaries, UNEP links atmospheric degradation to increased illness and deaths among humans. Among those most susceptible to these hazards, according to UNEP figures, are children and the elderly.

As you read, consider the following questions:

1. What two sources account for all air pollution, according to the authors?
2. What does UNEP list as "the most important indoor contaminants"?
3. What examples does UNEP provide to support its assertion that air pollution adversely affects human beings?

From the United Nations Environment Programme's State of the Environment (1972–1992) report *Saving Our Planet: Challenges and Hopes*, UNEP/GCSS.III/2 (Nairobi, Kenya: UNEP, 1992). Courtesy of the UNEP.

Atmospheric pollution is a major problem facing all nations of the world. Various chemicals are emitted into the air from both natural and man-made sources. Emissions from natural sources include those from living and non-living sources (e.g., plants, radiological decomposition, forest fires, volcanic eruptions and emissions from land and water). These emissions lead to a natural background concentration that varies according to the local source of emission and the prevailing weather conditions. People have caused air pollution since they learned how to use fire, but man-made air pollution (anthropogenic air pollution) has rapidly increased since industrialization began.

A Human Problem

Research over the past two decades has revealed that, in addition to the previously known common air pollutants (sulphur oxides, nitrogen oxides, particulate matter, hydrocarbons and carbon monoxide), many volatile organic compounds and trace metals are emitted into the atmosphere by human activities. Although our knowledge of the nature, quantity, physico-chemical behaviour and effects of air pollutants has greatly increased in recent years, more needs to be known about the fate and transformation of different pollutants and about their combined (synergistic) effects on human health and the environment.

World-wide, 99 million tonnes of sulphur oxides (SO_x), 68 million tonnes of nitrogen oxides (NO_x), 57 million tonnes of suspended particulate matter (SPM), and 177 million tonnes of carbon monoxide (CO) were released into the atmosphere in 1990 as a result of human activities, from stationary and mobile sources. The Organization for Economic Cooperation and Development (OECD) countries accounted for about 40 percent of the SO_x, about 52 percent of the NO_x, 71 percent of the CO, and for 23 percent of the SPM emitted into the global atmosphere, the rest of the world accounted for the remainder. Although the amount of SO_x emissions peaked in 1970 to a high of about 115 million tonnes, it dropped to 99 million tonnes in 1990 as a result of marked reduction in SO_x emissions in OECD countries. These reductions have been achieved mainly by stricter regulations of emissions, changes in energy structures and fuel prices and introduction of more efficient technologies. Between 1970 and 1990, SO_x emissions in the OECD region decreased from about 65 million tonnes to about 40 million tonnes. In contrast, SO_x emissions in the rest of the world increased from 48 million tonnes to 59 million tonnes over the same period. From 1970 to 1990, there were no marked changes in NO_x and SPM emissions. There was, however, a marked decrease in CO emissions in the OECD region, from 155 million tonnes to 125 million tonnes; in the rest of the world CO emissions increased from about 40 million tonnes in 1970 to 52 million tonnes in 1990, mainly due to an increase in automobile traffic.

In the past two decades, and especially in the 1980s, increasing attention has been given to the emission into the atmosphere of hundreds of trace compounds—organic and inorganic. Some 261 volatile organic chemicals (VOCs) have been detected in ambient air. In most cases, the concentrations are quite low, with a majority of chemicals at sub-part per billion by volume (ppbv) levels. Some of these VOCs are highly reactive, even at such low concentrations, and are suspected of playing a considerable role at least in the formation of photochemical oxidants. Another group of compounds that has received attention in recent years is trace metals, such as cadmium, mercury, zinc, copper, etc. Lead is the best studied of these metals. An estimated 80–90 percent of lead in ambient air derives from the combustion of leaded petrol.

Substandard Standards

A study of six cities has found that air pollution, even in areas that meet Federal air quality standards, can shorten people's lives. . . .

The study said mortality rates from lung cancer, lung disease and heart disease were 26 percent higher in Steubenville, Ohio, the most polluted area studied, than in Portage, Wisconsin, which was the least polluted.

The New York Times, December 9, 1993.

Because of growing concern about air pollution, programmes were initiated in some developed countries in the 1960s to monitor the common pollutants and assess changes in air quality. In 1973, the World Health Organization (WHO) set up a global programme to assist countries in operational air pollution monitoring. This project became a part of the United Nations Environment Programme's (UNEP) Global Environmental Monitoring System (GEMS) in 1976. Some 50 countries now participate in the GEMS/AIR monitoring project and data are obtained at approximately 175 sites in 75 cities, 25 of them in developing countries. Data from GEMS/AIR for the period from 1980–1984 indicate that of 54 cities, 27 have acceptable air quality (e.g., Auckland, Bucharest, Bangkok, Toronto, Munich) with sulphur dioxide concentrations below 40 micrograms/cubic metre (WHO established a range of 40–60 micrograms/cubic metre as a guideline for exposure to avoid increased risk of respiratory diseases). Eleven cities have marginal air quality (e.g., New York, Hong

Kong and London) with sulphur dioxide concentrations between 40 and 60 micrograms/cubic metre. The other 16 cities have unacceptable air quality (e.g., Rio de Janeiro, Paris and Madrid) with sulphur dioxide concentrations exceeding 60 micrograms/cubic metre. Data for 41 cities indicate that 8 of them have acceptable air quality with respect to SPM (e.g., Frankfurt, Copenhagen and Tokyo), with SPM concentrations below 60 micrograms/cubic metre (the WHO range is 60–90 micrograms/cubic metre). Ten cities have borderline concentrations of SPM, between 60 and 90 micrograms/cubic metre (e.g., Toronto, Houston and Sydney), and 23 cities have SPM concentrations exceeding 90 micrograms/cubic metre (e.g., Rio de Janeiro, Bangkok and Tehran). The extraordinary levels noted in some cities in developing countries can be partially explained by natural dust; other culprits include the black, particulate-laden smoke spewed out by diesel-fuelled vehicles, lacking even rudimentary pollution control. The GEMS/AIR assessment concluded that nearly 900 million people living in urban areas around the world are exposed to unhealthy levels of sulphur dioxide and more than one billion people are exposed to excessive levels of particulates. . . .

Urban Smog

Ozone and other photochemical oxidants, such as peroxyacetyl nitrate (PAN), are typically formed in the lower atmosphere from NO_x and hydrocarbon emissions in the presence of sunlight during stagnant, high-pressure weather conditions. This occurs most often during summertime, and leads to the well-known photochemical smog episodes, characterized by a thick layer of brown haze. Ozone concentrations in OECD countries, where time-series data are available, have not shown a clear trend; the principal reason is that they depend largely on prevailing weather conditions, which can change considerably from year to year. In many OECD countries, ozone levels exceed recommended standards. In the United States, the limit of exposure of 235 micrograms/cubic metre (for one hour/day as a maximum) is often exceeded, and it has been estimated that some 75 million people are exposed to higher levels of ozone.

Ozone has long been considered to be the oxidant that determines the air quality of an urban atmosphere. During the 1980s, however, atmospheric chemists identified hydrogen peroxide, a photochemical product in the air, as another oxidant that may significantly degrade air quality. Measurements of hydrogen peroxide carried out at various locations in Brazil, Canada, Europe, Japan and the United States show concentrations generally less than 10 parts per billion (ppb) by volume. No guidelines have yet been established for exposure to ambient hydrogen peroxide.

Air pollution is not only restricted to the outdoor environment.

Although indoor air pollution has been known since prehistoric times, and elevated concentrations of air pollutants continue to be a fact of life for people who live in impoverished areas and cook over open fires fuelled by charcoal, coal, wood, dung and agricultural residues, the problem of indoor air pollution has recently become a matter of concern. The expression "sick building syndrome" has been used to describe buildings in which the air causes a number of symptoms (e.g. eye, nose and throat irritation; mental fatigue; headache; nausea; dizziness: airway infection; sensation of dry mucous membranes, etc). Such symptoms have been epidemiologically related to sealed buildings, non-openable windows, tight-enclosure dwellings, increased temperature and dust levels, and passive cigarette smoking.

Indoor air pollution in residences, public buildings and offices is created for the most part by the occupants' activities and their use of appliances, power equipment and chemicals; by emissions from some structural or decorative material; by thermal factors; and by the penetration of outdoor pollutants. The most important indoor contaminants are tobacco smoke, radon decay products, formaldehyde, asbestos fibres, combustion products (such as NO_x, SO_x, CO, carbon dioxide and polycyclic aromatic hydrocarbons), and other chemicals arising from use in the household. WHO has indicated that several microbiological air contaminants are encountered in the indoor environment. These include molds and fungi, viruses, bacteria, algae, pollens, spores and their derivatives. Recently, more than 66 volatile organic chemicals have been identified in indoor air. Several studies have pointed out that many pollutants are more concentrated in the indoor environment than in the outdoor. Respirable particulate matter, NO_x, carbon dioxide, CO, formaldehyde and several other compounds and radon are higher indoors than outdoors. . . .

Impacts of Atmospheric Pollution

Air pollution affects human health, vegetation and various materials. The notorious sulphurous smog which occurred in London in 1952 and 1962 and in New York in 1953, 1963 and 1966 clearly demonstrated the link between excessive air pollution and mortality and morbidity. Such acute air pollution episodes occur from time to time in some urban areas. In January 1985, an air pollution episode occurred throughout western Europe. Near Amsterdam, the 24-hour average SPM and SO_x concentrations were each in the range of 200–250 micrograms/ cubic metre (much higher than the WHO guideline values). During the episode, several people were affected; pulmonary functions in children were 3 to 5 percent lower than normal. This dysfunction persisted for about 16 days after the episode. Athens is known for the frequent occurrence of such acute air pollution episodes. But even in the absence of such episodes, long-term exposure to air pollution can affect several susceptible groups (the elderly, children and those with respiratory and heart conditions). . . .

Local Pollution Is a Global Problem

Indoor air pollution has a number of effects. Reference has already been made to the sick building syndrome, which causes a substantial portion of disease and absenteeism from work or school. Recently, attention focused on the possible health hazards of radon emissions at home. In the United States, it has been found that the concentration of radon indoors is about 6 times higher than that outdoors and that the current annual mortality rate from lung cancer attributable to indoor radon exposure is about 16,000 cases. However, it was found that only 3 percent of this mortality occurred among individuals who never smoked tobacco. Thus, more than 90 percent of the lung cancer risk associated with radon could be controlled by eliminating smoking. The penetration of outdoor pollutants into buildings has also been a cause of concern. High ozone levels have been found in some museums and art galleries and there are fears that ozone—a highly reactive gas—could cause the fading of colours of art work. Several museums and art galleries have taken costly precautions to monitor ozone levels indoors and to shield paintings and other art work tightly.

Emissions from burning biomass fuels, especially in rural areas of developing countries, are a major source of indoor air pollution. The most important identified adverse effects are chronic obstructive pulmonary disease and nasopharyngeal cancer. When infants are exposed to such pollution, acute bronchitis and pneumonia occur because respiratory defenses are impaired. Emissions from biomass and coal burning at home contribute significantly to outdoor air pollution in some areas. It has been found that indoor emissions create a visible haze in certain parts of the Himalayas, which may have effects on visibility and on vegetation in that mountain ecosystem. . . .

Although it was thought that urban (and rural) air pollution problems were local problems, it has become increasingly evident that urban emissions lead to the regional and global distribution and deposition of pollutants. These scales are not isolated from one another, and solutions to problems in one may lead to new problems from another. For example, the use of tall stacks to disperse pollutants may abate local air pollution, but it causes regional and global distribution and deposition of primary pollutants and their reaction products. Therefore, in the past two decades it became evident that countries have to work in concert to reduce air pollution.

"Contaminated water has become commonplace in America."

America's Drinking Water Is Unsafe

Melissa Healy

Since becoming law, federal regulations designed to ensure the safety of America's drinking water have provoked debate. Many members of the Republican Party want these laws dismantled, contending that they generate needless expense and government red tape. Melissa Healy, a staff writer for the *Los Angeles Times*, argues that this action is misguided. She notes that many water districts throughout the United States fail to meet existing water quality standards and, as a result, threaten the health of Americans. Unacceptable levels of contaminants in drinking water, Healy writes, can cause serious ailments such as acute digestive disorders, long-term mental impairment, and sometimes even death.

As you read, consider the following questions:

1. Name a contaminant found in water and the specific health problems Healy attributes to it.
2. What legislation was enacted to ensure that water is safe to drink? What two claims do water treatment organizations make as their reasons for opposing it?
3. How extensive is contamination of drinking water in the United States, according to the information cited by Healy?

With the 104th Congress moving toward a major rewrite of laws governing the purity of the nation's drinking water, two environmental groups on June 1, 1995, released studies showing that nearly half of the nation's drinking water systems have exposed consumers to sickening contaminants such as lead, pesticides and the parasite cryptosporidium.

Contamination Is Commonplace

"Contaminated water has become commonplace in America," said Richard Wiles, president of the Environmental Working Group and co-author of a study titled "In the Drink." Congress' efforts to relax regulations, he declared, "might seem a good idea but not if it allows polluters to relieve themselves into our drinking water."

In two separate reports, the Environmental Working Group and the Natural Resources Defense Council concluded that the water supply of 45 million Americans was tainted between 1993 and 1995 with cryptosporidium—the microorganism that made 400,000 ill and killed more than 100 in Milwaukee in 1993—and that 53 million Americans during the same period have received tap water laced with levels of lead, pesticides and chlorine byproducts that exceed federal standards.

Both groups have lobbied aggressively to tighten existing laws and regulations governing the use of pesticides and have warned frequently of the public health dangers of industrial contaminants. In many cases, they have sought stricter regulation than that proposed by the Environmental Protection Agency (EPA).

The studies mark the environmental community's opening salvo in an effort to slow the drive of Republican lawmakers to roll back federal regulations.

Relaxing the Rules

The U.S. House of Representatives in mid-May of 1995 voted to reauthorize the Clean Water Act with major changes that would relax rules requiring industries, farms and cities to treat waste water before releasing it into the nation's lakes, rivers and oceans. And lawmakers are expected to begin redrafting the Safe Drinking Water Act, which establishes standards for tap water, in a bid to ease regulation of water treatment agencies.

Many water treatment organizations have lobbied hard for changes in the Safe Drinking Water Act, arguing that its purification, monitoring and enforcement provisions are costly to apply and frequently are not justified by the dangers of contaminants. But the authors of the studies contend that tainted water may be the cause of many illnesses, such as stomach distress or long-term loss of mental capacity, which are frequently not linked to water consumption.

Environmentalists have warned that relaxing both the Clean Water and the Safe Drinking Water acts would erode the cleanliness of the nation's waterways and put serious strain on the quality of drinking water.

Dirty Water

A list of U.S. cities with "drinking water to watch," according to the Environmental Working Group.

City	Violation
New York City	turbidity, coliform (feces)
Tucson, Ariz.	gross alpha radiation
Greenville, S.C.	coliform
Utica, N.Y.	coliform
Elizabeth, N.J.	coliform
New Port Richey, Fla.	coliform
Decatur, Ill.	nitrate, poor disinfection
Lansdale, Pa.	tetrachloroethylene, coliform
Joliet, Ill.	radium, gross alpha radiation, lead
Springbrook Twnsp., Pa.	poor disinfection, lead
Ft. Bragg, N.C.	trihalomethanes, lead
Altoona, Pa.	not filtering, coliform, poor disinfection
Rock Hill, S.C.	trihalomethanes, poor disinfection
Oak View, Calif.	turbidity
Bloomington, Ill.	nitrate
Shamokin, Pa.	poor disinfection, 1,1-dichloroethylene
Camden, N.J.	trichloroethylene, lead
Davis, Calif.	selenium
Danbury, Conn.	turbidity
Merchantville-Pennsauken, N.J.	carbon tetrachloride

Source: Environmental Working Group

Los Angeles Times, June 2, 1995.

While declaring that their conclusions do not warrant public panic, the authors of the two reports urged Americans to press lawmakers to oppose any loosening of existing laws. In the meantime, they urged Americans with compromised immune systems—the elderly, those undergoing chemotherapy and those who have AIDS—to take special care to avoid contaminants, choosing perhaps to boil their water before drinking it.

In Southern California, the Natural Resources Defense Council noted that cryptosporidium has been found in source water as well as in finished or filtered water of both the Metropolitan Water District, which serves more than 16 million customers, and the Los Angeles Department of Water and Power (DWP), which serves 3.6 million customers.

Jay Malinowski, spokesman for the Metropolitan Water District, confirmed that the agency did find cryptosporidium in two water samples taken on the same date in 1993 but noted that it has found none since. He said that the agency has stepped up its monitoring for the microorganism—a step that goes beyond state and federal regulations—but sees no reason to institute costly and aggressive new filtration processes.

Bruce Kuebler, director of water quality for the DWP, said that 16 of roughly 200 samples taken between January 1994 and June 1995 have yielded cryptosporidium. Although the concentration of the contaminant was "extremely low," he said, the DWP has taken pains to alert the Los Angeles County and state health departments and groups with suppressed immune systems, who appear to be particularly vulnerable to cryptosporidium.

The Environmental Working Group cited the Water Department of Pico Rivera as one that has had a number of violations of EPA standards. In 1993–94, Pico Rivera, which serves 35,000 customers, reported two instances of contamination by coliform bacteria, usually found in fecal matter.

Nationally, 1,172 water systems, serving 11.6 million people, reported contamination with fecal coliform, which causes gastrointestinal illnesses. Lead, which causes permanent loss in mental capacity in children, appeared to be the second most common water contaminant nationally. Violations of federal standards for lead were reported in 2,551 systems nationwide, serving 10.3 million people.

The environmental groups culled much of their data from water agencies' reports to the Environmental Protection Agency, but noted that many lawmakers are intent on discontinuing such reporting requirements, as well as public notification requirements of the Safe Drinking Water Act.

These lawmakers also want to relax monitoring requirements for a number of contaminants and allow small and medium-sized water systems to get waivers for purification requirements that are too costly.

> "Current law is the root cause of . . . lost confidence in our drinking water, not the quality of the water."

America's Drinking Water Is Safe

Jonathan Tolman

In the following viewpoint, Jonathan Tolman argues that government regulation of America's drinking water is excessive. Tolman criticizes the Environmental Protection Agency's (EPA) regulatory standards, contending they are impossible to achieve—both in practice, and in dollars. Furthermore, he asserts, these unnecessarily stringent standards create the impression that the nation's drinking water is unsafe, when in fact it is harmless. Tolman has served as a visiting fellow at the Alexis de Tocqueville Institution in Arlington, Virginia, and was previously an environmental analyst for former vice president Dan Quayle's Council on Competitiveness.

As you read, consider the following questions:

1. What, in Tolman's opinion, prevents the EPA from developing those regulations that are most urgently needed?
2. According to Tolman, what problem does federal water regulation create for local governments?
3. What important information do the media neglect when reporting on water safety, according to the author?

Jonathan Tolman, "EPA Regulates Against Unreal Risks to Drinking Water," *Human Events*, February 4, 1994. Reprinted with permission.

Wouldn't you know it, as soon as the North American Free Trade Agreement (NAFTA) passes you can't drink the water any more. In January 1994, Washington, D.C., residents became suddenly leery of something they normally take for granted, their drinking water.

Now some environmentalists and lawmakers, led by Representative Henry Waxman (D.-Calif.), are pointing to Washington's recent crisis as an example of inadequacies in our drinking water laws. They argue that what we need to do is tighten our drinking water standards in order to increase public confidence in our water supply.

Law and Water

In reality, the current law is the root cause of this lost confidence in our drinking water, not the quality of the water. Funding for drinking water programs has increased 47 percent since 1990. In addition to increased funding, the 10 years from 1984 through 1994 have seen an explosion in drinking water technology. Today, there are more sophisticated ways of analyzing, controlling and treating our drinking water. These two factors have created a drinking water system capable of efficiently delivering 40 million gallons of safe water to our homes every day.

What happened in Washington is a classic example of how the Safe Drinking Water Act itself is responsible for the loss of public confidence—and adding new regulations will only exacerbate the problem.

The Safe Drinking Water Act of 1986 requires the Environmental Protection Agency (EPA) to establish drinking water standards for 83 specific contaminants by 1989, and then every three years pick another 25 contaminants and set standards for them as well. Many of these standards apply to contaminants with theoretical cancer risks based on excessive doses fed to rats and EPA is a little behind schedule in setting standards for many of them. Since 1986 [and up through February 1994] they have set only 80 new standards.

Localities Pay for 95 Percent of Mandates

The problem is created when the local water agencies are required to meet these new standards. According to EPA, the total cost for complying with the new standards is $1.4 billion a year. The federal government, of course, generously offers grants to help pay for the new standards that they are imposing on local water agencies. The budget for the entire grant program in 1993 was $59 million, less than 5 percent of the cost for implementing the new standards.

This means that local governments have two choices. First they can try to come up with the other 95 percent of the funds to implement the new standards—which means either raising taxes, deficit spending or cutting services somewhere else. Or second, they can keep the same budget and try to squeeze better performance out of their current system and hope that they don't violate any of the new standards.

Washington, like many other cities, has chosen the latter option. In 1988, the budget for operating and maintaining the Washington Aqueduct was $14.6 million. In 1992 the budget was $16.2 million. This is an increase of about 2 percent a year, less than the rate of inflation. During this same five-year period, EPA created 66 new drinking water standards.

A Modern Myth

One of the modern myths about water is that natural sources are the best. Actually, few sources of water are found in a state that can be considered clean and usable as is. Virtually all water is treated to some degree. In fact, that's how society manages to support such large numbers of people living longer and longer without early mortality from disease. . . .

What many people forget is that natural water has impurities in it, too. These impurities may be higher than the levels found in ordinary tap water.

Common Sense, March/April 1994.

The most recent change in standards, and the one that caused the recent crisis, occurred in June 1993, when EPA changed the standard for turbidity. Turbidity is nothing more than a fancy word for cloudiness, and is harmless.

At 1 a.m. on Tuesday, Dec. 7, 1993, the Dalecarlia Water Treatment plant violated the new EPA standard for three hours. If this scenario had happened in May that same year, before the new standard was in place, there would have been no crisis. But because there was a new standard, the treatment plant had to inform EPA of the violation. This set off a bureaucratic chain reaction and suddenly more than a million people were drinking bottled water with funny sounding French names.

The water supply was no more dangerous than it had been six months ago. Water quality had not changed, only the standard had changed. And public confidence in the water supply was the biggest victim.

Media Ignore EPA's Changing Standards

The media do not headline a story when EPA changes a standard. Neither the media nor the public pay much attention to the obscure science of drinking water standards. So when the front-page headline suddenly reads, "EPA Officials Fear Contamination," the public automatically assumes that what has changed is the quality of their water.

This creates a vicious cycle of new standards, that can't be met because of insufficient funding, which leads to violations of the new standards, which leads to lower public confidence, which leads to calls for more regulation, which leads to new standards. . . .

The current drinking water law not only erodes public confidence, it also hinders the government's ability to respond to real threats. As it turns out, what EPA was really scared of in the Washington water crisis was not the harmless turbidity, but cryptosporidium, a parasite that made several hundred thousand residents in Milwaukee sick in the spring of 1993. For some reason, EPA felt that what was happening in Washington was "eerily" similar to what happened in Milwaukee.

During 1993 there were 336 violations of EPA's turbidity standard. During that same time, there was only one outbreak of cryptosporidium, the Milwaukee case. In fact, the water at the Milwaukee plant prior to the outbreak did not violate EPA's standard. The highest turbidity level in the Milwaukee plant at the time of contamination was well below EPA's current standard, and easily met the turbidity standard at the time. There is no apparent statistical link between violations of EPA's turbidity standard and outbreaks of cryptosporidium.

Some evidence suggests that the cryptosporidium in Milwaukee was simply a particularly virulent strain of the parasite. For example, high levels of cryptosporidium have been found in the Ohio, Missouri and Mississippi Rivers—as many as 480 parasites per liter. In spite of this, no outbreaks of cryptosporiosis have been reported in cities that use these rivers as their source of water.

EPA Standards Not Effective

If the Milwaukee outbreak is a valid example, it shows that it is all too possible for cryptosporidium to infect a water system without violating any current water standards. In fact, EPA appears to have a contaminant standard for just about everything except cryptosporidium. The agency hasn't even developed a method for detecting the parasite, let alone established how many parasites constitute an infective dose.

Since EPA has no official regulations for dealing with cryptosporidium, the agency's reaction to the turbidity violation is not surprising. The minute someone suggested that there just possibly might be some kind of virulent parasite in the water, which the agency had no regulation for dealing with, EPA played it safe, infinitely safe.

Why hasn't EPA come up with a standard for dealing with cryptosporidium? Basically, the agency hasn't had time to get around to it. The Safe Drinking Water Act requires EPA to spend all its time writing standards for the other contaminants that Congress specified in 1986.

The vast majority of these contaminants are obscure chemicals like the by-products of chlorine disinfection. Researchers have discovered that these chemicals, known as trihalomethanes and haloacetic acids, when force-fed to rats in large doses can cause cancer. The Safe Drinking Water Act then requires EPA to come up with standards for these chemicals.

Real Money vs. Theoretical Risk

In effect, it is a form of regulation based on theoretical cancer deaths, not on actual cancer statistics. In many cases EPA standards are forcing cities to spend their real money to prevent EPA's theoretical cancer risk. This chemical-of-the-day approach often wastes government resources on setting standards for chemicals that pose no verifiable risk, instead of focusing the resources on real health problems.

This is why the U.S. Conference of Mayors and the National Governors Association have called for an end to unfunded federal mandates. Local politicians know why the people in their cities, counties or states are dying. Despite how bad it might taste, how many people died last year in Washington from drinking the water? How many died of heart disease? How many were shot?

There is only so much money in any city's budget. If members of Congress want to spend real money on theoretical problems, then they should spend their own.

"It is the land, not oceans, that faces the greatest threat."

Ocean Pollution Is Manageable

R.L. Swanson, J.R. Schubel, and A.S. West-Valle

Whenever news of oil spills or medical waste dumping in the ocean reaches the public, people grow alarmed and angry. While much of this concern is justifiable, argue R.L. Swanson, J.R. Schubel, and A.S. West-Valle, the authors of the following viewpoint, land is the resource most crucial to human survival, and its preservation must take priority over that of the ocean. They maintain that the ocean has shown itself to be remarkably resilient to pollution by humans, especially where cleanup programs have been enacted, and may prove the more environmentally sound alternative for waste disposal. Swanson is an adjunct professor at the State University of New York (SUNY) Marine Sciences Research Center and director of the Waste Management Institute. Schubel serves as dean and director of the SUNY Marine Sciences Research Center. West-Valle works as an editorial associate at the Waste Management Institute.

As you read, consider the following questions:

1. According to the authors, why should environmental protection of the land take priority over that of the oceans?
2. How do the authors propose adapting existing environmental legislation for the future?
3. What do the authors advocate as an alternative to prohibiting ocean waste disposal?

Excerpted from "Are Oceans Being Overprotected from Pollution?" by R.L. Swanson, J.R. Schubel, and A.S. West-Valle, *Forum for Applied Research and Public Policy*, Spring 1994. Reprinted with permission.

Indignation and concern are appropriate responses to the problem of coastal degradation. However, society must not be misled when the media leaves the impression that the oceans are hopelessly degraded by pollution.

The United Nations (U.N.) advisory group on marine pollution states without equivocation that "no area of the ocean and none of its resources appear to be irrevocably damaged."

Given current estimates of global population growth during the next century—from about 6 billion in 2000 to about 10 billion in 2050—it is the land, not oceans, that faces the greatest threat. Moreover, nearly 100 percent of the world's drinking water and 90 percent of its protein comes from the land. If society must make choices about environmental protection—and it must—the land must be protected first.

Nevertheless, the persistent environmental stresses that the marine environment faces due to population growth and uncontrolled coastal development do not bode well for the health of the ocean. Increased coastal contamination is inevitable.

Legislating for the Ocean's Health

Concern for estuarine and coastal waters is long-standing; legislative and regulatory initiatives have sought to ameliorate some of the problems. The Federal Water Pollution Control Act (Clean Water Act) of 1948 and its numerous amendments through 1987, the National Environmental Policy Act of 1969, the Marine Protection Research and Sanctuaries Act of 1972 (Ocean Dumping Act), the Coastal Zone Management Act of 1972, the National Ocean Pollution Planning Act of 1978, the Marine Plastic Pollution Research and Control Act of 1987, the Ocean Dumping Ban Act of 1988, and the Medical Waste Tracking Act of 1988 all focus on the need to improve the environmental quality of the nation's coastal waters.

The court of world public opinion also has been influential. The Japanese, for example, will be phasing out or modifying their current practice of using drift fishing nets that add to marine litter and kill or maim marine organisms.

The environmental challenges facing the world's oceans are simple to state but difficult to address. The nations of the world must:
- Reduce wastes.
- Effectively manage the wastes that remain.
- Understand the impacts of residual wastes on air, land, and water.

The last challenge should serve as the basis of the next generation of environmental policy initiatives. The benefits derived from legislation and regulations that specify permissible concentrations of contaminants probably have been reached. The effectiveness of new laws must be improved. Better strategies for managing marine pollution problems that cut across the global commons—of which the oceans are a part—are needed.

Legislative incentives to reduce pollution should be encouraged. Such measures would ease the burden of finding acceptable means to handle remaining wastes. Environmental awareness is growing in the industrial sector, in part, because it is good business. Reuse of materials, development of secondary materials, and use of less polluting materials all are having positive effects, even if such endeavors remain limited.

Promising Prospects

Planetary restoration, which promises to be the great environmental endeavor of the next few centuries, has begun in the seas. Urban estuaries are surprisingly resilient, and biologists have learned a good deal about how to recall them to life.

Ken Brower, *Omni*, April 1989.

Industry is using more of what was once considered scrap. Chemical exchange programs are being developed. Scrap plastic resins are now recycled largely because of public demand. Plastic scrap made into plastic lumber, in fact, may be advantageous to marine industries because wood, concrete, and steel deteriorate so quickly in marine environments.

Restrictions have proven beneficial in many instances. For example, the banning of polychlorinated biphenyls (PCBs) in the United States dramatically reduced concentrations of these compounds in coastal marine sediments and some marine organisms. In fact, the National Oceanic and Atmospheric Administration (NOAA) reports that PCB concentrations did not increase in any of its national mussel-watch stations between 1986 and 1988.

Backed into a Terrestrial Waste Corner

Marine systems are resilient. These ecosystems appear to have the capacity to rejuvenate, despite assaults on their integrity through high concentrations of pollutants and other human-induced insults. Polluted areas around several southern California sewage outfalls have improved as a result of more stringent regulations and the relocation of primary sewage treatment effluent discharge. Concentrations of PCBs have decreased among species of fish and shellfish by 100-fold since 1969. On the Atlantic coast, there is evidence, based on changes in species and the rising population of invertebrates, that a 12-mile sewage sludge site near New York City is recovering only two years after reform measures were implemented to reverse six decades of dumping.

Oceans have a tremendous capacity to assimilate, dilute, neutralize, and disperse wastes. In terms of human health, land-based disposal technologies, such as composting or incineration, may not be more environmentally sound than ocean disposal—particularly in areas where groundwater and air emissions are a concern. In brief, highly restrictive legislation, such as the Ocean Dumping Ban Act of 1988 or the Clean Water Act as presently integrated, may not be in the nation's best long-term environmental interest.

Instead of prohibiting the use of oceans as a media for handling wastes, more attention should be given to pretreating or preventing contaminants from entering the waste stream. This strategy would produce cleaner effluents and sludges, thus mitigating the environmental impact of residual wastes in all environments.

Laws and regulations should acknowledge that ocean and estuarine processes vary considerably and that this variability can be used to ameliorate the impacts of pollutant discharges. For example, secondary sewage treatment may not be required or represent the best strategy for the marine environment in all coastal areas even though it is the law of the land.

Public sentiment and, as a result, public policy are easily swayed by media attention. The focus is currently on the environmental hazards of air pollution and its impact on climate. With ocean disposal now legally off limits, we may have backed ourselves into a terrestrial waste disposal corner because we have been forced to turn to the least regulated media: the land.

Neutralized or stabilized residual wastes (wastes that have been physically and/or chemically altered to reduce adverse interactions with the environment) can be disposed of in the ocean with little negative impact, thus avoiding risks to drinking water and saving valuable terrestrial space for other uses. Offshore islands, for example, can be used to isolate waste-management activities that may not be desirable near large population centers.

The Need to Act Globally

René Dubos encouraged society to "think globally but act locally" in dealing with environmental pollution. But scale is important, too. If not appropriately planned on a sufficiently large scale, a localized approach could result in piecemeal, short-sighted, and often conflicting laws, rules, and regulations.

There are positive indications of a willingness of nations to implement agreements to protect the oceans. An example is the ban on the disposal of plastics at sea through the International Maritime Organization's MARPOL V. Other international bodies, such as the United Nations Environment Programme, the International Oceanographic Commission, and the International Council for Exploration of the Seas, also are serving as forums to protect oceans and their resources and help define and bring into the political process acceptable uses of the ocean.

The appropriate role for oceans in a waste-management strategy or the appropriate scale and means to effectively manage increased coastal development and its attendant pollution has not yet been identified. An assessment of these issues should be undertaken, building on the excellent studies by the National Advisory Committee on Oceans and Atmosphere and the Office of Technology Assessment. Important issues that should be included in such an assessment are:

- The waste-management role of oceans, *vis-a-vis* land and air.
- Pollution-related impacts of projected population growth in coastal areas.
- The variability of coastal and ocean waters to accommodate waste.
- Current knowledge about the effects of ocean pollution.
- Use of the ocean as a repository for stabilized wastes.
- An assessment of ocean use to accommodate long-lived unneutralized or unstabilized contaminants.

"Rapid population growth, industrial expansion, [and] rising consumption . . . are causing levels of marine pollution . . . that constitute a global threat."

Ocean Pollution Is a Serious Problem

Peter Weber

In the following viewpoint, Peter Weber argues that the earth's oceans are in peril from human-generated waste and pollution. Weber contends that clean-up efforts are difficult and only partially successful in the face of continuing encroachment and overdevelopment of coastlines. He describes humanity's impact on the ocean as the equivalent of declaring war. Weber is a research associate for Worldwatch Institute, an environmental research organization in Washington, D.C.

As you read, consider the following questions:

1. Why are coastal cities a threat to the marine environment, in Weber's opinion?
2. To what extent has the flow of nutrients and sediments into the ocean increased, according to the author?
3. Name one example that Weber uses to illustrate the decline of marine life resulting from the pollution of the ocean.

Excerpted from "Oceans in Peril" by Peter Weber, *E Magazine*, May/June 1994. Reprinted with permission.

Since the beginning of life on Earth, the oceans have been the ecological keel of the biosphere. The marine environment, from the brackish waters where rivers flow into the sea to the deepest depths, constitutes roughly 90 percent of the world's inhabited space. The oceans cover nearly 71 percent of the Earth's surface, and their deepest trenches plunge lower below sea level than Mount Everest climbs above it. They hold 97 percent of the water on Earth, more than 10,000 times as much water as all the world's freshwater lakes and rivers combined.

The Source of Life

The oceans' seminal contribution to the planet was life itself. The first living organisms on Earth are thought to be bacteria that developed in the depths of the seas some four billion years ago. Not only are they the evolutionary ancestors of us all, but they created the oxygen-rich atmosphere—a key to our existence—as a by-product of their photosynthesis. Even today the oceans are a foundation of global climate, and they are home to a unique array of species, many of which cannot be found on land. Remarkably, deep sea dredges indicate that the ocean floor may contain as many species as the world's tropical rainforests. Many of the species brought to the surface cannot be identified because they have never been seen before, and are unlikely to be caught again. Scientists are increasingly turning to the sea because of its unique biological diversity. They have derived anti-leukemia drugs from sea sponges, bone graft material from corals, diagnostic chemicals from red algae, anti-infection compounds from shark skin and many more useful agents.

Time and evolution have distanced us from our oceanic origins, but we still bear the traces of our saltwater heritage in our blood. We have an almost universal fascination with the timeless procession of waves, the smell of salt water, the call of seabirds, the sheer scale of the sea. From the vantage point of a beach or a coastal cliff, the oceans look limitless and unchanged from the way they appeared thousands of years ago. Throughout most of human history, we have seen only this view, and our governments have made few, if any, attempts to protect the marine environment. Today, however, with technologies that allow us to penetrate the salt water depths, we have discovered that the oceans, too, are vulnerable to the unsustainable trends that degrade the environment on land. Rapid population growth, industrial expansion, rising consumption and persistent poverty are causing levels of marine pollution, habitat destruction and depletion of marine life that constitute a global threat to the marine environment.

If we were to declare war against the oceans, the most destructive strategy would be to target the coasts, the regions of most highly concentrated biological activity. Tragically, that is what we are already doing—not by deliberate attack, of course, but through overcrowding of coastal areas and unsustainable economic development. Here is where agricultural and urban waste pours in from the land, smoggy clouds pour out their contaminants, ships flush their tanks, and cities bulldoze wetlands to extend their land seaward. Over half the people in the world now live within 100 kilometers of the coast, while coastal cities make up nine of the 10 largest cities and over two-thirds of the top 50 in the world. As these cities continue to grow, developers drain wetlands that once served to trap nutrients, sediments and toxins, so that runoff from construction, city streets, sewage plants and industrial facilities now flows unimpeded into coastal waters.

The Signs of Crisis

The last decade has not been easy on the world's oceans, the largest and seemingly least vulnerable resources on Earth. . . .

Signs of a crisis are hard to miss:

Nearly half of Europe's harbour seals were destroyed by viral infections in the 1980's. Ulcerative mycosis—classified only in 1978—has wiped out fleets of food fish like the Atlantic menhaden. The sea's most delicate and commanding coral reefs, from the Florida Keys to Australia's Great Barrier Reef, have begun to die; some slowly as logging silt chokes off sunlight; others suddenly as divers squirt in cyanide to stun fish for aquariums. . . .

Even the essential chemical balance at many river mouths has changed as nitrogen-rich fertilizers and acid rain have increased while fresh water is siphoned off by dams and development. As the water's composition changes, so does the sea's intricate food chain—starting with plankton, and ending with people.

Michael Specter, *The New York Times*, May 3, 1992.

The world is dotted with cities that have degraded their coastal habitats. San Francisco Bay, the largest estuary in the western United States, which has shrunk by 60 percent with land reclamation in the last 140 years, is overrun by alien species, and can no longer support commercial fishing. Off the Palos Verdes peninsula just south of Los Angeles, a city sewage plant contributed to the progressive elimination of 7.8 square kilometers of kelp forest as it increased its discharge 20-fold between 1928 and 1966. Sludge laced with toxins and heavy metals covered 95 percent of the former kelp bed. But the most extensive source of habitat destruction along the coasts actually occurs in rural areas, partly because poor people quit scarce farm land to try fishing. Shrimp farmers in the Asian and South American tropics, for instance, have cleared extensive tracts of mangroves for holding ponds; they now produce 20 percent of the world's shrimp supply. But about half of the world's salt marshes and mangrove swamps have been cleared, drained, diked or filled. Five to 10 percent of the world's coral reefs have been ruined by pollution and destruction—another 60 percent could be lost in the next 20 to 40 years. Even beaches are endangered, with 70 percent eroding worldwide.

Stemming the Flow of Pollutants

The flow of nutrients into the oceans has at least doubled since prehistoric times, and sediments have nearly tripled due to human activity. Together, nutrients and sediments have become pollutants that degrade estuaries and coastal waters by blocking sunlight, suffocating fish and coastal habitats, and carrying pathogens and toxins. They have contributed to "red tides," blooms of algae that release deadly levels of toxins into the surrounding waters. In Japan's Seto Inland Sea, the number of red tides increased from 40 in 1965 to more than 300 in 1973. Three years later, after the Japanese authorities had introduced controls to limit the influx of nutrients, they began to decline in frequency. In other areas, however, the tides continue. The poisons released by them kill mass quantities of fish and can weaken or kill people through direct exposure or contaminated seafood.

Roughly one-half of the nutrients entering coastal waters comes from the ubiquitous problem of sewage and runoff from coastal cities and farms, but the other half, surprisingly, comes from inland. In the eastern United States, for instance, the Chesapeake Bay has been overwhelmed by nutrients from distant sources. Farmers contribute one-third and air pollution one-quarter of the nitrogen pollution that has crippled this estuary, the largest in the United States, and once one of the most productive in the world. The oyster catch in the Chesapeake has fallen from 20,000 tons in the 1950s to under 3,000 tons in the late 1980s, partly because of pollution. When Europeans first came to the United States, oysters in the Chesapeake could filter all the water in the Bay in two weeks. Now, because there are so few, they take more than a year.

But the Chesapeake also offers an example of what can be done to try to curb coastal pollution. This estuary is fed by over 150 tributary rivers and streams spread over six states and the District of Columbia. Under the Chesapeake Bay Agreement of 1987, the District of Columbia, Virginia, Maryland and Pennsylvania have until the year 2000 to reduce nutrients by 40 percent by controlling the discharge of toxins and increasing wetland area. So far, they have had mixed success. Seagrass area has increased by 57 percent, the Potomac is much cleaner, and phosphorus levels are down by 20 percent. But runoff from agriculture has increased, the population in the region continues to grow rapidly, development on the Bay coastline continues and, as a result, the load of nitrogen nutrients entering the Bay continues to increase. This experience demonstrates the difficulty of implementing such programs in a single country, much less worldwide. . . .

Working Together

The oceans ultimately belong to the world, and people and nations must work together to protect them. . . .

We can no longer afford to act as if the oceans are limitless or unalterable. The complex links between land and sea may make the task of protecting the oceans for future generations seem daunting, if not impossible. But it is precisely because of these links—because the oceans touch the lives of us all—that we cannot ignore their health if we are to protect our own place on the planet.

Periodical Bibliography

The following articles have been selected to supplement the diverse views presented in this chapter. Addresses are provided for periodicals not indexed in the *Readers' Guide to Periodical Literature*, the *Alternative Press Index*, or the *Social Sciences Index*.

Chemecology	"Clean Water: Now the Point Is Nonpoints," May 1994. Available from the Chemical Manufacturers Association, 2501 M St. NW, Washington, DC 20037.
Adel Darwish	"Arid Waters," *Our Planet*, vol. 7, no. 3, 1995. Available from United Nations Environment Programme, PO Box 30552, Nairobi, Kenya.
George Guthrie	"Breathing Easy?" *World & I*, April 1995. Available from 3600 New York Ave. NE, Washington, DC 20002.
Don Hinrichsen	"Sea Change," *Amicus Journal*, Spring 1995.
Joyce Price	"Gas Is Greener, but Smog Safer," *Insight*, April 24, 1995. Available from 3600 New York Ave. NE, Washington, DC 20002.
Marc Ross	"Forum: Why Cars Aren't as Clean as We Think," *Technology Review*, February/March 1994.
J.A. Savage	"The Road Warriors: Utilities and Automakers Square Off on Alternative Fuel Vehicles," *Business and Society Review*, Winter 1994. Available from 25-13 Old Kings Highway N., Suite 107, Darien, CT 06820.
Kenneth Silber	"Blowing Their Stacks over Pollution Rights," *Insight*, July 12, 1993.
Kirk R. Smith	"Taking the True Measure of Air Pollution," *EPA Journal*, October/December 1993. Available from Waterside Mall, 401 M St. SW, Washington, DC 20460.
Steven T. Taylor	"The Great Smog Conspiracy 25 Years Later: Lessons for the Future," *Public Citizen*, Fall 1994.
Andrew Weil	"Protect Yourself from Air Pollution," *Natural Health*, January/February 1994. Available from PO Box 57320, Boulder, CO 80322-7320.

Is the American Lifestyle Bad for the Environment?

The Environment

Chapter Preface

A cowboy named Clay Overson, a Death Valley prospector known as Sparrow, and the oldest trees in the world share something in common. The Desert Protection Act of 1994 shields their natural habitat—the California desert regions of Mojave and Death Valley—and their presence there as integral to its landscape. U.S. senator Dianne Feinstein, who spearheaded the legislation, says she chose to extend the act's protection to cattlemen like Clay because they are part of "our Western heritage."

Elsewhere in the West, the goals of environmentalists and ranchers lack this amiability. Environmentalists consider ranchers, as Christopher Christie terms it, "the greatest environmental threat" facing the western range—hardly something worth preserving. Activists argue that ranchers destroy many native grasses and other plant species critical to wildlife by routinely overgrazing their herds. According to Christie, 19 percent of the animals and 22 percent of the plants on the endangered species list are there as a direct result of this practice. Overgrazing also leads to soil erosion, environmentalists contend, which damages rivers and streams and causes destructive flooding.

Western leaders and ranchers argue that the presence of ranchers has benefited the environment. Ranchers maintain that they have built fences, installed windmills, and developed stock-watering tanks at their own expense—all of which benefit wildlife as well as their own herds. They claim that they take pride in their stewardship of the land and that their efforts have dramatically improved the range lands over the last sixty years. "Anybody making a living on a ranch would be dumber than a post to ruin the land," insists Warner Glenn, a third-generation cattleman in Arizona.

Environmentalists press for the protection of critical habitats through grazing restrictions; ranchers fear this could mean losing the only way of life their families have known for generations. Meanwhile, in the new Mojave preserve the grazing once restricted to one small park is now allowed throughout much of the protected desert lands—which have been granted official wilderness status to prevent any development whatsoever, even road building. How effective this compromise will be remains unknown.

At the center of the debate over the western rangeland lies the question of whether Americans must change their traditional—and contemporary—way of life or risk permanently destroying the environment. The argument continues as authors explore this question in the following chapter.

"From global warming to species extinction, we consumers bear a huge responsibility for the ills of the earth."

The American Consumer Is Bad for the Environment

Alan Thein Durning

According to Alan Thein Durning, the United States is birthplace to the consumer society—a way of life that threatens to deplete global natural resources. Durning, director of Northwest Environment Watch and former senior researcher at Worldwatch Institute, organizations that promote environmental conservation, contests the idea that economic health and personal fulfillment are dependent upon consumerism. He argues that if the children of future generations are to enjoy the abundance of nature as the current generation has, people must seek an alternative to the consumer lifestyle. The following viewpoint was adapted from Durning's book *How Much Is Enough? The Consumer Society and the Future of the Earth* and appeared in *Sierra* magazine.

As you read, consider the following questions:

1. Why is spiritual impoverishment a predictable result of consumer culture, according to Durning?
2. What justifications for embracing consumerism does Durning relate?
3. Name three of the author's examples of how services can be provided without depending on "high resource consumption."

Excerpted from "Long on Things, Short on Time" by Alan Thein Durning, *Sierra*, January/February 1993. Reprinted by permission of the author.

The consumer society was born in the United States in the 1920s. Economists and business executives, concerned that the output of mass production might go unsold when people's natural desires for food, clothing, and shelter were satisfied, began pushing mass consumption as the key to continued economic expansion. As brand names became household words, packaged and processed foods made their widespread debut, and the automobile assumed its place at the center of the culture, the "democratization of consumption" became the unspoken goal of American economic policy.

The Consumer Society

Since then, the consumer society has moved far beyond U.S. borders, expanding to Western Europe and Japan by the 1960s. Yet far outpacing growth of the consumer class itself—the 20 percent of the world's people who earn 64 percent of world income—is the spread of its underlying cultural orientation, consumerism. That term, writes British economist Paul Ekins, refers to the belief that "the possession and use of an increasing number and variety of goods and services is the principal cultural aspiration and the surest perceived route to personal happiness, social status and national success." But even as, over a few short generations, more than a billion of the world's people have become car drivers, television watchers, mall shoppers, and throwaway buyers, social scientists have found striking evidence that high-consumption societies have not achieved satisfaction. The consumer society fails to deliver on its promise of fulfillment through material comforts because human desires are insatiable, human needs are socially defined, and (perhaps most critically) the real sources of personal happiness are elsewhere. "The conditions of life which really make a difference to happiness," writes psychologist Michael Argyle, "are . . . social relations, work, and leisure. And the establishment of a satisfying state of affairs in these spheres does not depend much on wealth, either absolute or relative."

The consumer society, it seems, has succeeded mostly in impoverishing us. This impoverishment is profoundly spiritual, but it is also demonstrably environmental. Our way of life entails an enormous and continuous dependence on the very commodities that are most damaging to the earth to produce: energy, chemicals, metals, and paper. By drawing on resources far and near, we consumers—though our numbers are concentrated primarily in North America, Western Europe, Japan, Australia, Hong Kong, Singapore, and the oil sheikdoms of the Middle East—cast an ecological shadow over wide regions of the world. Our appetite for wood and minerals, for example, motivates the roadbuilders who open tropical rainforests to poor settlers, resulting in the slash-and-burn forest clearing that is condemning countless species to extinction. A blouse in a Japanese boutique may come from Indonesian oil wells by way of petrochemical plants and textile mills in Singapore and assembly factories in Bangladesh. Likewise, an automobile in a German showroom bearing the logo of an American-owned corporation typically contains parts manufactured in a dozen or more countries, and raw materials that originated in a dozen others. From global warming to species extinction, we consumers bear a huge responsibility for the ills of the earth.

Yet our consumption too seldom receives the attention of those concerned about the fate of the planet. Technological change and population stabilization—as reasonable as it is to emphasize their importance—must be complemented by a reduction of our material wants. Human desires will *still* overrun the biosphere unless they shift from material to nonmaterial ends. The ability of the earth to support billions of human beings depends on whether we continue to equate consumption with fulfillment. At the same time, we need to challenge the received wisdom that consumption must be pursued—regardless of its human and environmental effects—as a matter of national policy in order to keep ourselves employed.

Consume-or-Decline Thinking

The latter assumption runs deep. Broadcast news programs cover retail districts in the holiday season as if they were scenes of national significance, offering commentary on shoppers' readiness to buy. When recession hit the United States in mid-1990, everyone from the president on down began begging loyal Americans to spend.

The reasoning behind such entreaties sounds impeccable: if no one buys, no one sells, and if no one sells, no one works. Thus, in the consumer economy—where two-thirds of gross national product consists of consumer expenditures—everything from fortunes on the stock market to national economic policies hinges on surveys of "consumer confidence" and "intentions to buy." If this consume-or-decline view is right, then lowering our consumption on purpose, individually and collectively, would be self-destructive. Cutting our driving in half, for example, would throw half the gasoline station attendants out of work, along with half the car mechanics, auto workers, auto-insurance agents, and car-financing specialists. The shock of those layoffs would cause a chain reaction of additional job losses that could end in a repeat of the Great Depression.

Mainstream economists paint a similarly nightmarish scenario for developing countries. The industrial countries, they solemnly intone, are the locomotive of the world economy. Contracting demand in industrial countries would leave the citizens of impoverished lands stranded in destitution. Having gambled everything on consumers' endlessly growing appetite for their exported raw materials, developing nations would fall into irreversible decline. In this view, failing to increase consumer-class intake of raw materials is a crime against the 42 poorest nations—what the United Nations calls the least-developed countries—because they depend on commodity exports for more than 60 percent of their foreign earnings.

The consume-or-decline argument contains a grain of truth. The global economy is indeed structured primarily to feed the consumer lifestyle of the world's affluent fifth, and shifting from high to low consumption would shake that structure to its core. It would require legions of workers to change jobs, entire continents to reconfigure their industrial bases, and enterprises of all scales to transform their operations. It would, worst of all, entail painful dislocation for thousands of families and communities.

Unsettling Changes

But consider the alternative: continuing to pillage and poison the earth would guarantee not only the same misfortunes but worse. Fishers will be left idle if water pollution and overharvesting decimate fisheries. Farmers will abandon their fields if recurrent drought kills their crops and animals. Loggers will have little to do if the forests are destroyed by air pollution, acid rain, clearcutting, and shifting climatic zones. Carmakers and homebuilders will not find many buyers if people must spend most of their earnings on scarce food supplies. Business, in short, will not do well on a dying planet. In this light, the admittedly unsettling changes to the global economy so gloomily defended against by the consume-or-decline school are no more an argument against lowering consumption than job losses in the weapons industry are an argument against peace.

The contention that the world's poor cannot afford for us consumers to live on less is debatable at best. Although many developing countries are integrated into the world economy as suppliers of raw materials, this situation puts them in a dependent status that their leaders have decried for decades. Furthermore, the trickle-down effects of the growing consumer economy have proved a disappointing source of economic stimulus. Indeed, the most notable consequence so far has been to create enclaves of world-class consumers in every nation. These elites profit mightily from the exports of natural resources from the global South to the global North. But the world's poor have gained little beyond devastated homelands. Ending poverty, as innumerable experts confirm, depends on aggressive national campaigns for basic health, education, and family planning, on broad-based, labor-intensive development schemes in rural areas, on the mobilizing efforts of grassroots organizations, and on the existence of responsive local and national governments. To finance all these, economic policies must promote innovation, reward success, and allow markets to work efficiently. Ending poverty does not necessarily depend on the production of bulky, low-value goods destined for export to the consumer class. More beneficial by far would be world-trade rules written to make commodity prices reflect more of the ecological costs of production.

An Economy of Permanence

The transition to an economy of permanence will be challenging, but perhaps less so than the consume-or-decline argument suggests, because that line of thought rests on three arguable assumptions: that consumption of economic services is immutably bonded to consumption of physical resources in the economy; that employment is equally bonded to flows of physical resources; and that only one model of employment—40 hours a week, all year round—is viable, even though that model requires more hours devoted to work than have been committed by most civilizations in history.

Physical flows of natural resources would slow radically in an economy of permanence, while the money value of the services that people enjoy might fall little. The crucial distinction is between commodities and the services people use those commodities to get. For example, nobody wants telephone books or newspapers for their own sake; rather, we want access to the information they contain. In an economy of permanence, that information might be available to us for much the same price on electronic readers. That would enable us to consult the same texts but eliminate most paper manufacturing and the associated pollution.

Likewise, people do not want cars as such; they buy them to gain ready access to a variety of facilities and locations. Good town planning and public transportation could provide that access equally well. In every sector of the economy, from housing to food, this distinction between means (physical goods) and ends (services) helps lay bare the vast opportunities to disconnect high resource consumption from the quality of life.

High Labor, Low Environmental Impact

The total amount of work done in an economy of permanence may decrease little, because the most ecologically damaging products and forms of consumption also usually generate the fewest jobs. Indeed, there is a striking correspondence between high labor intensity and low environmental impact. Repairing existing products, for example, uses more labor and fewer resources than manufacturing new ones. Railway systems employ more people but fewer natural resources than comparable fleets of cars do. Improving energy efficiency would employ more people than would boosting energy production. And recycling programs employ more people than waste incinerators or landfills do.

Consuming Until Nothing Is Left

If we continue . . . to consume the world until there's no more to consume, then there's going to come a day, sure as hell, when our children or their children or their children's children are going to look back on us—on you and me—and say to themselves, "My God, what kind of monsters *were* these people?"

Daniel Quinn, *Utne Reader*, September/October 1993.

Were the consumer class to substitute local foods for grain-fed meat and packaged fare, to switch from cars to bikes and buses, and to replace throwaways with durable goods, labor-intensive industries would benefit greatly. Still, on balance, the amount of paid work done might decrease, because low-impact industries would probably expand less than high-impact industries contracted. To cope fairly with slackening job markets, societies in transition to low consumption would (among other initiatives) have to provide laborers in high-impact fields with sufficient job retraining, offer adequate unemployment compensation to smooth the process, and reduce the number of hours each person works. Fortunately, as far as this last-named strategy is concerned, most of us consumers work more than we wish to anyway—a topic we will examine in greater depth shortly.

Nature-Blind Economics

Governments will also find themselves addressing the employment question when they tackle—as they must—a radical reorientation of their prevailing tax and subsidy policies, many of which promote the worst kinds of consumption. Most nations favor their auto, energy, mining, timber, and grain-fed livestock industries with a long list of tax writeoffs and direct subsidies. The United States virtually gives away minerals on federal land, builds logging roads into national forests at taxpayer expense, and sells irrigation water in the arid West at a loss. France massively subsidizes its nuclear-power complex, Russia its oil industry, the United Kingdom its auto drivers, the Canadian province of Quebec its aluminum smelters, and Japan its feed-grain growers.

Beyond financial transfers and biased policies are the implicit subsidies of the nature-blind economic accounting systems that governments use. Coal and oil are not priced to reflect the damage their production and combustion cause to human health and natural ecosystems. Pulp and paper are not priced to reflect the habitat destroyed and water poisoned in their production. Scores of products—from toxic chemicals to excessive packaging—cost the earth more than their price tags reveal.

If goods were priced to reflect something closer to their full environmental costs, the market would help guide consumers toward lower resource consumption. Disposables and packaging, for instance, would rise in price relative to less-packaged goods; local unprocessed food would fall in price relative to prepared products trucked from far away. If legislators shifted the tax burden from labor to resources, companies would swiftly move to trim resource use as environmental taxes rose, and hire more people as income taxes fell.

Already, environmental and taxpayer groups in many nations single out egregious subsidies and tax shelters as targets for reform. But they commonly lose the big battles, overwhelmed by the political clout of the billion-dollar industries that doggedly defend the status quo. Every battle lost demonstrates the difficulty and the urgency of mobilizing more members of the consumer class in support of prices that tell the ecological truth.

Enough Stuff

As previously suggested, an important approach to a low-consumption, sustainable future is for us consumers to release ourselves from the strictures of full-time work. Many of us find ourselves agreeing with American industrial designer William Stumpf, who says, "We've got enough stuff. We need more time."

Although fulfilling work and adequate leisure are both key determinants of human contentment, the balance in the consumer society tilts too far toward work. Working hours in industrial societies, although far below their peak during the Industrial Revolution, remain high by historical standards. Japanese and Americans are especially overworked. Europeans have been trading part of their pay raises for additional leisure time since 1950, but Americans and Japanese have not. In Germany and France, the average hours worked per week have gone from 44 and 38 hours respectively in 1950 to 31 hours in 1989, with much of the decline reflecting annual vacation leaves spanning four to eight weeks. In Japan, weekly hours have gone from 44 to 41. In the United States, meanwhile, the workweek declined slightly from 1950 to 1970, but has actually increased since then. Americans work 38 hours a week, on average, and have added an entire month's worth of work to their schedule since 1970.

Harvard University economist Juliet Schor writes in *The*

Overworked American: "Since 1948, the level of productivity of the U.S. worker has more than doubled. In other words, we could now produce our 1948 standard of living in less than half the time. Every time productivity increases, we are presented with the possibility of either more free time or more money. We could have chosen the four-hour day. Or a working year of six months. Or every worker in the United States could now be taking every other year off from work—with pay." Instead, Americans work the same hours and earn twice the money. To check whether that choice reflected the will of American workers, Schor delved into the arcane field of labor-market economics and, having surveyed dozens of studies, concluded that it did not. Workers in all the core regions of the consumer society express—either in opinion surveys or in collective-bargaining positions—a strong desire for additional leisure time and a willingness to trade pay increases for it. They also report that they do not have that option: They can take a job or leave it, but they cannot take it for fewer hours a day. Part-time work, furthermore, is in general less skilled, less interesting, and less well paid because it lacks fringe benefits such as retirement and insurance plans. So most of us are left with the choice of good full-time jobs or bad part-time ones.

Trading Pay for Time Off

In an earlier era, cynics said workers would squander free time on drinking and gambling. But when the W.K. Kellogg Company shortened its workday from eight hours to six during the Great Depression, community initiatives proliferated. Contemporary observer Henry Goddard Leach noticed "a lot of gardening and community beautification. . . . Athletics and hobbies were booming. . . . Libraries [were] well patronized . . . and the mental background of these fortunate workers . . . [was] becoming richer."

Mounting pressure for more time instead of more money is evident in things like the campaign of some 240 U.S. labor, women's, and children's organizations for the right to take time off for family and medical purposes. The coalition, chaired by the Women's Legal Defense Fund, pushed a bill through Congress in 1991 that was vetoed twice by President Bush. Similarly, for two decades unions such as Service Employees International have been supporting voluntary work-time reduction programs for workers who want free time instead of money. The Service Employees won such a program temporarily for California state government workers in the 1970s, and later prevailed permanently for New York state government employees.

More recently, interest has surged in flexible work arrangements such as job sharing, particularly among women stretched thin by the "double day" of career and family. So far, unfortunately, American men have not joined women in pushing for flexible or reduced hours, perhaps because our culture disapproves of men who don't want to work full-time. Meanwhile, in Japan, where *karoshi* (death from overwork) kills perhaps 10,000 people a year, young workers are displaying a newly disapproving attitude toward overtime, pressuring colleagues to leave the office at the end of the scheduled day. The Japanese government plans to switch the country from a six-day workweek to a five-day one by early in the next century. In Europe, too, unions continue to press for additional time off.

The Simple Life

No one can say yet how strong this preference for free time over extra consumption is. In theory, if everyone consistently chose free time over additional money, normal gains in labor productivity would cut consumer-class working hours in half by 2020, giving us abundant time for personal development and for family and community activities.

A culture of permanence will not come quickly. Much depends on whether we, among the richest fifth of the world's people, having fully met our material needs, can turn to nonmaterial sources of fulfillment. Can we—who have defined the tangible goals of world development—now craft a new way of life at once simpler and more satisfying? Having invented the automobile and airplane, can we return to bicycles, buses, and trains? Having introduced the junk-food diet, can we nourish ourselves on wholesome fare that is locally produced? Having pioneered sprawl and malls, can we recreate human-scale settlements where commerce is an adjunct to civic life rather than its purpose?

If our grandchildren are to inherit a planet as bounteous and beautiful as we have enjoyed, we in the consumer class must—without surrendering the quest for advanced, clean technology—eat, travel, and use energy and materials more like those on the middle rung of the world's economic ladder. If we can learn to do so, we might find ourselves happier as well.

"*Radical environmentalism is undoubtedly [why Worldwatch] Institute endorses . . . devastating sacrifices of (other) people's lives and welfare.*"

The American Consumer Is Being Unfairly Vilified

Stephanie Moussalli

In the following viewpoint, Stephanie Moussalli argues against the views of the Worldwatch Institute, including the premise that Americans are among the world's most wasteful and polluting consumers. Moreover, she contends that people living within societies identified as "sustainable" suffer in misery from a lack of modern technology and basic human freedoms, and would eagerly choose a lifestyle similar to that of the average American if allowed. Such is the case with Bhutan, Moussalli notes, the example cited by Worldwatch staffer Christopher Flavin as the ideal of "sustainable development." Moussalli is an instructor of history and political science at Pensacola Junior College in Florida.

As you read, consider the following questions:

1. On what basis does Moussalli challenge the assumption that Americans' use of the automobile poses a significant threat to the global environment?
2. Why does Moussalli consider Worldwatch staffer Christopher Flavin a hypocrite?
3. What does the author identify as the flaws in the "sustainable development" of Bhutan?

Stephanie Moussalli, "Worldwatch Watch," *National Review*, April 18, 1994; ©1994 by National Review, Inc., 150 E. 35th St., New York, NY 10016. Reprinted by permission.

The Worldwatch Institute exists, in its own words, to "save the planet," a miracle it is accomplishing by raising the public consciousness of the need "to create an environmentally sustainable global economy." The Institute, led by Lester R. Brown, is one of the more influential and reputable of environmentalist policy organizations. It produces the widely quoted *State of the World*, an "annual checkup of our global environment," which is used, says the Institute, as a text in five hundred university courses around the country.

So a few years ago I began reading the bi-monthly *World Watch* in the hope of learning the strongest of the environmentalist arguments. Instead, a review of the last four years of this staff-written magazine—which "seek[s] to change the course of history" and which the Institute inflicts on the publics of five other countries in addition to the United States—reveals a utopian ecosocialist world view quite as hostile to human welfare as traditional socialism has ever been.

Attacking Consumerism

The magazine despises consumerism, for example. "The planet is in danger" from the excessive consumption of the developed world, whose populations of "car drivers, beef eaters, soda drinkers, and throwaway consumers" are devoted, in the tolerant opinion of researcher Alan Durning, to "greed, envy, and avarice." Advertising causes a great deal of this consumerism, substituting "Coca-Colonization" for "clean drinking water" in the Third World, and "third cars and fourth televisions" for "environmental literacy [and] social and sexual equality" in the developed world. One of the few ads Durning likes portrays North Americans as "a gigantic animated pig" and is entitled "High on the Hog." Instead of malls, Durning wants "communities [to] turn main streets into walking zones where local artisans and farmers display their products while artists, musicians, and theater troupes perform. . . ."

Of course, circumstances alter cases. After Durning vilified the new Minneapolis Mall of America as "a monument to higher consumption," in the September/October 1992 issue, the staff made a rueful announcement in the next issue that Worldwatch's retirement-fund investors are the majority owners of the mall. Since then, the magazine has dropped the subject of malls and investment funds.

The "Scourge" of the Automobile

World Watch also worries a lot about automobiles, "the scourge of the late twentieth-century civilization." This is partly because they produce "one-fourth of the greenhouse gases that now threaten the stability of the global atmosphere," but even without global warming, cars are unsustainable because cities do not have "the physical space" to expand parking. *World Watch* writers often use words idiosyncratically; since more than 90 per cent of the greenhouse effect is produced by water vapor, and natural sources account for about 90 per cent of the atmospheric CO_2, "one-fourth of the greenhouse gases" must mean something other than what it says. "Physical space" is another creatively used term, since all the cities and roads in the U.S. combined cover only about 4 per cent of the country's surface, but let it go; the point is that *World Watch* approves of Al Gore's "modest suggestion that the pollution-spewing internal-combustion engine be phased out in another quarter-century."

What should replace cars? The Institute approves of mass transit, solar-powered vehicles, and walking, but what it really loves is bicycles, "the most efficient form of transport ever invented." After all, many cities find bikes "more effective than gridlocked squad cars in chasing criminals." Tanzania uses them to deliver mail, and Malawi uses bicycle-ambulances. In fact, in May, when long daylight hours render Washington, D.C., streets relatively safe, "as many as fifteen" of *World Watch*'s own employees "come to work on their bikes."

The Semantics of Ecosocialism

Ecosocialist semantics sometimes confuse the unwary reader. I, for instance, got quite excited about a proposal for maritime "property rights" in the November/December 1992 issue, until I realized the author meant governments should divvy up the high-seas commons among themselves. Danish farmers are "making extra money" from compost-produced methane, declared another piece, and, as an enthusiastic gardener, I saw the dollars rolling in, until the next sentence announced this "project" was made possible by "government subsidies." China should adopt "real market prices" for energy, announced one writer, and I thought the *World Watch* staff must have learned some economics, until I read that China "has already made a promising move on this front, having planned to triple the price of oil sold domestically."

The magazine occasionally makes its semantic preference explicit, as in a September/October 1990 feature article in which Sandra Postel asks that Gross National Product (GNP) be redefined to take "environmental damage" and "human welfare" into account. This, she argues, will cause decision-makers to understand "that at some point growth begins to cost more than it's worth." However, one example she gives—"Anyone who has lost a favorite park to a new housing subdivision knows that not all economic growth enhances the quality of life"—reveals her lofty concern with conceptual economics to be little more than the hostility to newcomers typical of urban homeowners who have lost their free scenic views to the development of a nearby property.

"Stark Raving Mad"

In any event, Miss Postel's belief that word-fiddling will alter real marginal costs and benefits edges over into what the inimitable Petr Beckmann [a pro-nuclear energy scientist who dubbed the environmental movement the "Green Church" and saw it as antagonistic to technology and progress] used to call the "stark raving mad" area, and *World Watch* spends a lot of time there. For example:

• "There is a growing belief among scientists that no level of radiation exposure is safe" (May/June '91). Right. Since blood is radioactive, what we need is a National Vein Drain Day.

• If governments would only end "the hatreds, racism, and economic and environmental problems that cause people to flee their homes," then refugee problems would solve themselves (November/December '92). Lester Brown for President in '96!

• "Despite being only one among the millions of animal species, human beings control up to 40 per cent of the earth's potential photosynthetic productivity [PPP] from land areas" (March/April '93). I give up, and most of the rest of you had better, too, if we're going to reduce human energy use to one ten-millionth of PPP.

World Watch also opposes bananas, gold mining, fast foods, and meat. It disapproves of beltways, automation, dry cleaning, the spread of conifers in Scotland, and "tree farms masquerading as reforestation projects." On the other hand, the Institute wants to save the world's glaciers and mountains, especially the Himalayas. Actually, I thought the staff had finally slipped its leash on this one until, flipping through a recent *National Geographic*, I saw a nostalgic and moving plea to save the glaciers in southern British Columbia, and realized *World Watch* is just following the eco-fads of the Nineties. And the money; one hundred different agencies currently have people in Nepal to save the Himalayas.

On to Shangri-La

So what kind of world would get the Worldwatch Institute's seal of sustainability? The answer appeared in a July/August 1990 article with the cover title, "Bhutan, Paradise Preserved."

"Bhutan?" I squawked. How dare anyone describe that wretched little Himalayan despotism as Eden? As the imagined strains of "Brigadoon" [a 1968 musical about a village that is idyllic, unaffected by time, and remote from reality] wafted through the distance, I sat down to read.

Bhutan is a tiny mountain kingdom south of Tibet where most of the population lives in pre-nineteenth-century conditions. In

1990–91, the per-capita GNP was $180; out of every 100 people, only about 15 could read; life expectancy was around 48 years. On the Physical Quality of Life Index, a scale of 100 on which the United States scored 95 in the mid 1980s, Bhutan got 25.

To author Christopher Flavin, this is paradise. The inside title of the article is "Last Road to Shangri-La." Shangri-La, he tells us, is "the rarest of exceptions: a 'poor' nation where most needs are met, the traditional culture has been preserved, and the natural environment is still largely intact."

Mr. Flavin had just spent four weeks advising the Bhutanese government on how to avoid "the single-minded approach to economic growth" which has brought to "many countries . . . not only falling water tables and choking pollution, but a decade of declining incomes and swelling bellies." The process by which economic growth causes declining incomes is apparently so well understood by *World Watch* readers as to need no explanation. What Mr. Flavin does explain is how "Bhutan has already crossed the perceptual threshold so critical, and so elusive in other countries, to a sustainable society."

It hasn't been easy; the country has had to deal with the "emergence of an industrial sector," "rapid growth in population," and "consumerism." "New business interests," too, "can . . . lead to political corruption of the sort that has stripped the resources and impoverished the people of so many nations."

Handling "Tough Choices"

But the government of Bhutan can handle the "tough choices." It outlawed most private logging in 1979, more recently forbade "a major marble quarry and hydropower project," decided to keep "at least 60 per cent of the country covered in forest," and has put aside some of the best land in the country as "extensive wildlife preserves" and an "elephant corridor." In the opinion of King Jigme Singye Wangchuk, says Flavin, "industrial development and wealth accumulation should be secondary to the goal of sustainability," so the king has declined to invest "huge quantities of capital in counterproductive infrastructure and industries." In 1989, he even protected cultural sustainability by dismantling all his subjects' television antennas. After all, as Flavin points out, when "people are exposed to the 'things' that are available to those who live in industrial societies they often want to have the same . . ."

"Sustainable development," in short, means "no development."

Compassion for the Peasantry

In any event, "wealth accumulation" has not troubled most Bhutanese, the vast majority of whom live in hamlets of four or five huts connected to the rest of the country solely by mountainous mule tracks. Flavin is not without compassion for these peasants, many of whom never encounter outsiders except for "the occasional development worker." The peasants certainly deserve compassion—they drop like flies from tuberculosis and malaria, and have one of the highest infant-mortality rates in the world.

Courtesy of Paul Taylor, ©1990.

Flavin's solution? Rooftop solar panels, to be paid for by international aid. Solar power is so diffuse that hut-sized panels can produce only a tiny amount of electricity, but Flavin estimates it's enough for "two efficient mini-fluorescent bulbs and a radio—the main uses for power in rural Bhutan."

Of course, if some of the country's incredibly abundant hydropower were developed, and if King Jigme Singye would relax his tyrannical grip on the economy, the Bhutanese would soon be taking care of themselves, and then they might do anything. They might actually prefer power tools, cars, and polyester to their alternatives, even if it meant that suburbia came to Shangri-La. Some of them might become development advisors just like Flavin, and the UN would fly them to exotic parts of the world, and at home they would listen to CDs, and on holidays they'd

90

ship their packages in styrofoam.

What hypocrisy! If Flavin really believed that the life led by the Bhutanese peasant—ameliorated by two fluorescent light bulbs and a little more health care, to be sure—is Shangri-La, he would be living such a life himself. He might even join the actual Bhutanese peasants behind their oxen, except that another thing not allowed in Shangri-La is immigration. No problem; there are places right here in the belly of the capitalist beast where the committed may live the same life. Isolated forests, primitive agriculture, hundreds of acres of land with just one or two people living in a hut (with solar panels)—Flavinesque sustainability is legal and in fact easily available in America.

Flavin doesn't live that way. He has the choice, and he chooses styrofoam.

It's hard to avoid concluding that there are people who love coercive planning a little more than they love the natural world, and far more than they love the Bhutanese. King Jigme Singye, for one, is no stranger to the tools of tyranny. In January of 1990, Bhutan rated 5.5 on the Freedom House [a U.S.-based political and civil liberties organization which monitors human rights worldwide] political- and civil-liberty scale of 1 to 7. This was about the same as the USSR in its last months. Since Flavin's visit, it has deteriorated to 7, a category shared only by such countries as China, Cuba, and North Korea. Freedom House does not issue such low rankings lightly. Only governments practicing "extreme oppression" and "political terror" qualify for a 6 to 7.

Fearless Totalitarianism

Christopher Flavin, however, is not one to panic in the face of totalitarianism. After all, he is involved with projects like the World Commission on Environment and Development, which is making plans "for the entire planet into the distant future." In any event, people don't place too highly in his ranking of values, as he indicated in a casual remark that "[m]aintaining the health and happiness of the population is as essential to protecting the natural environment as any program to control logging. . . ."

Why such antipathy to people and freedom at Worldwatch? The adolescent millenarianism typical of radical environmentalism is undoubtedly one reason the Institute endorses such devastating sacrifices of (other) people's lives and welfare. Then, too, there is the sheer glamor of power. Insisting on democracy wouldn't elicit many invitations to spend "four weeks . . . advising the Royal Government" with "a private meeting" with the king thrown in. And this king, handsome and young, exotically garbed but fluent in English, married to four sisters but frequently to be found on his basketball court, is practically a fairy-tale prince. Western writers frequently compare his regime to Camelot, and refer to its magical qualities. The seduction of Flavin must have been a matter of routine to Jigme Singye.

Oppression Masquerades as Preservation

Flavin's sort of Shangri-La ought to be rated "X" and reserved for consenting adults only. But the Bhutanese haven't consented to this life on the cutting edge of ecosocialism. If I were Bhutanese, and saw a child die from a sudden fever that antibiotics could have cured if it hadn't taken two days to walk to a clinic, or watched my husband grow old prematurely behind a team of oxen and a friend, unmarriageable because of physical handicaps, grow sour for lack of modern job opportunities, and knew that one of the causes of all this was Worldwatch's apocalyptic chic, I think that I might even come to regard Christopher Flavin and company as oppressors and exploiters instead of paradise preservers.

Stable democracies like the United States mute the influence of fanatic movements, so that even an ecosocialist Vice President, for example, must constantly compromise. In such countries, Worldwatch-type policies afflict primarily that relatively small percentage of the public which is poor, while only making the middle class less comfortable, rather than desperate. But in some parts of the Third World, Worldwatch's genteel passion weds the naked violence of government in its worst state, and begets frequent misery and occasional horror. This is roughly the same allocation of misfortune between the free and unfree worlds that traditional socialism made. Perhaps we need a new version of the [now defunct] Committee for the Free World, which, through 1990, devoted itself to delegitimizing leftist fellow-travelers in the West. Certainly something is needed to prevent Worldwatch from changing "the course of history."

"While we say we like . . . old-style town life, we accept suburban development patterns . . . that destroy community by obliging us to drive everywhere."

Suburbia Is Harmful to the Environment

Neal Peirce

Urban sprawl has characterized the growth of American cities since World War II, when mass production entered the building industry, and resulted in a proliferation of suburban subdivisions. Syndicated columnist Neal Peirce contends that, among numerous other environmentally damaging features, the design of these neighborhoods promotes an increased dependence on the automobile, even when workplaces are nearby. The cumulative costs of this inefficient land development, Peirce maintains, are paid in ecological degradation, urban blight, increased pollution, and loss of community.

As you read, consider the following questions:

1. What, in Peirce's opinion, is inefficient about the way American land is developed?
2. What are the costs, according to the author, of developing "edge cities"?
3. What does Peirce present as an alternative to current development patterns?

Neal Peirce, "Americans: Conservationists or Champion Land Hogs?" *Liberal Opinion Week*, September 13, 1993; ©1993 Washington Post Writers Group. Reprinted with permission.

Check in almost any region you like and you discover that Americans *say* they want diverse, colorful, people-friendly, well-used downtowns and neighborhoods.

A rush of "Vision 2000" or "2020" plans have been produced by broad-based citizen committees—some acting independently, some with government support—from Seattle to Orlando, Indianapolis to Louisville, Chattanooga to Tempe, Arizona. Almost all place high value on historic preservation, lively streets, public transit accessibility.

Nor, do we seem to like the idea of endless suburban sprawl. Our vision statements endorse preservation of open space in and around our regions.

Yet there's a deep chasm between what Americans say we want and what, as consumers choosing places to live and work, we actually do.

Champion Land Hogs

We may indeed be the champion land hogs of history. Our urban areas devour land four to eight times faster than their population grows. The New York area's population increase over the past 25 years has been only 5 percent, but the developed land has increased by 61 percent, devouring nearly 25 percent of the region's forests and farmland.

And while we say we like—even romanticize—old-style town life, we accept suburban development patterns of rigidly compartmentalized, single-purpose land uses that destroy community by obliging us to drive everywhere.

We say we value public transit, but we stick to our private cars. During the 1980s, while mass transit systems languished, the number of Americans driving their cars all alone to work soared by 22 million, or 35 percent, several million more than the 18 million workers added to the national work force.

We call for preserving the open countryside, but we rush to the new subdivisions and "edge cities." Virtually all European "citistates" have set up and enforce urban growth boundaries. But not us. Only Oregon enforces truly strict growth boundaries around its cities.

The Stages of American Sprawl

Real estate expert Christopher Leinberger traces multiple stages of U.S. commercial development. From the old downtowns, we began to branch out in the '60s to close-in commercial centers like Bala Cynwyd near Philadelphia and the northern reaches of Phoenix's seemingly interminable Central Avenue. Then, from the mid '70s through the '80s, development exploded into such mega-developments as Tysons Corner, Virginia, in the Washington suburbs, King of Prussia, Pennsylvania, Denver's

94

Tech Center and Bellevue across Lake Washington from Seattle.

Yet another wave began in the late '80s—to such distant development centers as Houston's 290 corridor, Loudoun County, Virginia (an hour's congested drive from Washington), and Mesa outside of Phoenix.

In the meantime, our inner cities have suffered immense disinvestment and have lost hundreds of thousands of jobs.

'Earth? We used to go there but it's been ruined by developers.'

And what next? When the recession truly ends, will investment in our downtowns and older suburbs pick up again? Are we ready to reject exploitative sprawl?

Probably not, say most of the experts. "Now that so many jobs have moved to the fringe," predicts Leinberger, "new housing can go another 20 miles out. There will be a leapfrog effect."

Metropolitan Cleveland provides a typical example. As long as most jobs were in or near downtown Cleveland, a 30-minute commute limited people to fairly close-by suburbs in Cuyahoga County.

But now, with the interstates complete and so-called "edge

city" commercial centers sprung up around the regional periphery, the 30-minute commuting range has moved far beyond Cuyahoga County. And potential new suburbs show up as an eerie blob on the planners' maps, spreading far out into still-rural areas of Lorain, Medina, Summit, Portage, Geauga and Lake Counties.

"Edge cities," reports *EcoCity Cleveland Journal*, "not only make it possible for suburbia to chew up more woods and farmlands. They also promote more commuting across suburbs, more congestion on suburban and rural roads, more energy consumption and pollution, and a more dispersed population that cannot be served by mass transit."

To that, one must add the gruesome social cost as new job locations proceed farther and farther out into the countryside. Some inner-city folks might have found a way to get to jobs in a Country Club Plaza beside Kansas City, or Towson near Baltimore, or Los Angeles's Century City. But it's virtually impossible for them to get to such distant places as Ontario (34 miles from Los Angeles), or Hoffman Estates—the spot 37 miles from Chicago's Loop to which Sears escaped, jettisoning practically all its Chicago work force.

An Unconcerned Corporate America

Our land use practices threaten social conflagration in our cities. But corporate America doesn't seem to care. Not a single major corporate relocation in America in the last four years, notes Leinberger, "has gone anywhere except the absolute metropolitan fringe."

In the boardrooms, the distant "campus-like" settings may seem to make sense—the land, after all, is cheap and easily assembled, and it turns out that the executives, with uncanny regularity, live nearby.

But the public costs—despair in the inner cities, environmental degradation, undermining of older neighborhoods and suburbs—are frightening. Romantic visions about our communities need to be transformed into tough political action that sets meaningful growth boundaries around our burgeoning citistates—and insists that *all* developers and corporations honor them.

"The same number of cars in a smaller area [means] more congestion and the increased pollution associated with it."

Suburbia Is Not Harmful to the Environment

Randal O'Toole

Many urban planners argue that low-density land uses, such as suburbs, are environmentally unsound. Economist Randal O'Toole disputes this view. His own low-density hometown of Oak Grove, Oregon, he avows, has little pollution-producing automobile traffic, and residents regularly practice environmentally safe activities such as walking, bicycling, and raising their own food. O'Toole maintains that high-density urban growth, promoted for its purportedly superior efficiency and environmental benefits, actually creates more pollution than the suburbs. O'Toole is a contributing editor to *Liberty* magazine, a libertarian bimonthly publication produced by the Invisible Hand Foundation in Port Townsend, Washington.

As you read, consider the following questions:

1. What trade-off does O'Toole predict if Americans leave the suburbs for city dwelling?
2. Name two specific neotraditional zoning measures opposed by O'Toole. How does he say these changes would alter the character of his neighborhood?
3. What explanation does the author offer as the basis for zoning laws? How does this contradict the neotraditional movement?

Excerpted from "The Battle of Oak Grove" by Randal O'Toole, *Liberty*, September 1995. Reprinted with permission.

Soviet-style social engineering is alive and well in the United States. And the most avid practitioners are not in Washington, D.C., but our own communities—specifically, in our city and county planning offices. A new breed of social engineer—the "new urbanist"—has taken root there. Convinced that the automobile is the most monstrous invention ever devised, the new urbanists have concocted plans that would actually *increase* congestion and air pollution. According to their Orwellian reasoning, cities are congested with cars because they don't have enough people, so their solution is to cram more people in. Streets, they say, are busy because they are too wide, so their solution is to make streets narrower. And people drive a lot because their homes are so ugly, so planners want to impose strict design requirements on all new construction.

Illogical Land Use

I first encountered this logic in 1977, when I was trying to understand federal land agencies such as the Forest Service. Someone convinced me that I could find answers at the University of Oregon's Department of Urban and Regional Planning, so I moved from my hometown of Portland to Eugene to enter the masters program.

Most students in the program took no classes outside the department, but for some reason, I signed up for a class in urban economics. Fortunately, it was taught by Ed Whitelaw, a soon-to-be-well-known Northwest economist who helped us build simple models of an urban economy, test those models against reality, make the models more complex, and test them again. I was awestruck by the power of the economic way of thinking.

At that time, the legislature had recently enacted a law requiring cities to draw urban growth boundaries around themselves. Development outside these boundaries would be forbidden, or at least discouraged, until all available sites inside the boundaries were fully developed. The theory was that this would protect prime farmland. But our models clearly showed that since the measure resulted in roughly the same number of cars in a smaller area, the tradeoff would be more congestion and the increased pollution associated with it (cars pollute more in stop-and-go traffic).

Soon after developing this model, one of my urban planning classes considered the same question: *What effect would an urban growth boundary have on congestion and pollution?* The budding planners unanimously reasoned that, with a smaller urban area, more people would walk or use mass transit, reducing congestion and pollution. I tried to explain the economic model in 25 words or less, but convinced no one. I was particularly shocked to find that the two professors in the class agreed with the other students and considered my answer wrong.

That day I decided I was an economist, not a planner. During the next two years I crammed as many economics courses as I could take. Over the next ten years I applied the economic way of reasoning to the Forest Service (and, in 1988, published my findings in a book titled *Reforming the Forest Service*). During that time, I had little contact with urban planners, except for the times I served on hiring committees for various environmental organizations. Recently graduated planners were often attracted to such organizations, and I had the pleasure of automatically rejecting any candidate with a degree in urban planning.

In 1989, I moved back to the Portland area and settled in an unincorporated suburb known as Oak Grove. Unlike most suburbs, Oak Grove has a distinguished history stretching back to 1893, when the world's first electric interurban rail line was built between Portland and Oregon City. Wealthy Portlanders built grand houses on large estates along the line, and someone decided to call a small retail area next to one of the stations "Oak Grove."

Sounding Worse than It Really Is

Since the end of World War II, urban and suburban sprawl has consumed about 1 million acres of land a year. That sounds bad, but cities still cover just 3.1 percent of the total land area of the continental United States. In fact, the amount of land newly dedicated to parks and wilderness grew twice as fast as urbanized areas over the same period. This doesn't include the national parks, private forestland, and rangelands, which together account for 65 percent of total land area.

Ed Rubenstein, *National Review*, February 6, 1995.

Today, most of those large houses remain, but many of the estates have been subdivided. Still, most families live on lots of a quarter-acre or more. The block I live on is 20 acres, ten times the size of a normal city block. At a third of an acre, my lot is one of the smallest on the block. The temptation to subdivide further is tempered by the fact that the zoning requires a minimum lot size of 10,000 square feet (slightly less than a quarter-acre).

Such a low density means there is little auto traffic. When I walk with my dog around the neighborhood, I meet many of my neighbors who are also out walking. Since the houses were built one-by-one over a 100-year period, each is unique, with many made of native rock by a pioneer family of stonemasons. Many people take advantage of their large yards to grow flowers and vegetables, and a few own small livestock—poultry, goats, even a donkey.

The interurban railway is long gone, but several blocks away is Highway 99, the first four-lane road ever built in Oregon. Today it has developed into what many people scornfully refer to as a "strip development," though my neighbors and I find the numerous supermarkets, variety stores, and specialty shops to be convenient with competitive prices. And frankly, we're glad to keep most of the commercial areas some ways distant from our homes.

Suburbia Lost

I've spent more than five years here blissfully unaware of the new urbanism. Then, in April of 1994, a neighbor slipped a note into my mailbox suggesting that I should attend a county planning meeting. Ordinarily, I might have ignored it, but the note had an urgent tone that made me feel I should go.

At the meeting, I learned that county planners had been working for six months with "my neighbors" on a "transportation and growth management plan." The purpose of the plan, they claimed, was to give people more opportunities to walk and ride their bicycles than we have. This seemed peculiar since no one I know has ever felt hesitant to walk or bicycle around the neighborhood.

Then they showed us a map of the plan. The block I lived on and several nearby were to be rezoned for a 5,000-square-foot minimum lot size. The block across the street plus many others would be rezoned for multi-family dwellings with up to 24 units per acre. We were assured that this "densification," as the planners called it, was for our own good and that it would encourage walking and discourage cars.

To give people a place to walk to, a significant chunk of our neighborhood would be "mixed use," with stores and other businesses located a few steps from residences. In particular, planners hoped that many of the multi-family dwellings would be three stories high (two stories is the current limit) with businesses occupying the street floor.

When my neighbors and I asked about the reasoning behind these ideas, planners responded with totally circuitous logic. Why did planners want to densify our community? "Because densification is part of the neotraditional concept." What is neotraditionalism? "Neotraditionalism is a planning concept that calls for densification.". . .

Urban Relapse

After the meeting, I took some time to catch up on recent planning literature. Neotraditionalism turns out to be the brainchild of a California architect named Peter Calthorpe and a Florida husband-wife architect team named Andres Duany and Elizabeth Plater-Zyberk. Calthorpe has designed a neotraditional community near Sacramento and Duany and Plater-Zyberk have one near Miami.

These architects decided to model their plans after communities where people mainly walked, bicycled, or rode mass transit rather than drove. Based on their studies of such communities, they decided that people would reduce their use of cars if they lived in multi-family dwellings or in houses on small lots with tiny front yards and garages in back, and if their homes were close to grocery stores and other shops.

What communities did they study to reach these conclusions? Why, American communities of the 1920s. Americans had few cars back then, and they didn't drive much, so, the architects reasoned, if they designed communities like those of the 1920s, the people who live in them today won't drive much. (I am not making this up.) "Urban planning reached a level of competence in the 1920s that was absolutely mind-boggling," claims Duany.

"Zoning with an Attitude"

Duany, Plater-Zyberk, and Calthorpe are confusing cause and effect. People lived in "neotraditional" communities in the 1920s because they didn't have cars. But that doesn't mean that people will abandon their cars if we force them to live in such communities again.

Were those cities of the 1920s so wonderful? Not to the planners of the day. Contrary to Duany's claim, those urban areas weren't planned—they just happened. It was the unplanned mixture of uses to which today's planners want to return that inspired zoning in the first place.

About the same time I began to question neotraditionalism, *Newsweek* magazine featured a cover story on the new urbanism. Such wonderful ideas as small lot sizes, design codes, and corner grocery stores, the magazine gushed, would cure the suburban blues.

The 13-page article barely admitted that anyone might not appreciate the architects' latest schemes. But it did quote an urban planner who questioned whether anyone "really wants to recreate the social ambiance of an eighteenth-century village" and an economist who thought architects had a "strange conceit . . . that people ought to live in what they design." The magazine dismissed these comments as the natterings of free-marketeers.

But most of the problems that new urbanists complain about aren't the fault of the "free market." Instead, they are the legacy of previous generations of planners. Duany's ideal cities of the 1920s were almost totally unplanned—and planners hated them.

By contrast, the large suburban lots and strip developments that we have today are the direct result of the zoning ordinances that planners imposed on cities in the 1940s and 1950s. Mixed uses were supposed to be unhealthy, so planners separated them. Small lot sizes were unappealing, so planners zoned for minimum lots of 7,000, 10,000, or more square feet.

Though many planners blame urban blight on the automobile, some planners recognize that zoning is the true cause. Says Randall Arendt, a planning professor at the University of Massachusetts, zoning "is why America looks the way it does. The law is the major problem with the development pattern."

Zoning is not a cure that is worse than the disease. It *is* the disease. Planners have become nostalgic for a time before most of them were born—a time that people in their profession said was so bad that it could only be fixed by zoning. And how are we supposed to get back to this wonderful time? More zoning.

But this time we will have "zoning with an attitude." Says Plater-Zyberk: "Most zoning codes are *proscriptive*—they just try to prevent things from happening without offering a vision of how things should be." For example, a traditional zoning plan might contain a provision calling for 5,000-square-foot building lots at a *minimum*, allowing people to use 10,000- or 20,000-foot building lots if they want.

In contrast, says Plater-Zyberk, neotraditional zoning is *"prescriptive.* We want the streets to feel and act a certain way." That is to say, neotraditional zoning might call for building lots with a *maximum* size of 5,000-square feet. Such prescriptive zoning is used in many places, including parts of the Portland area. Multi-family zones forbid construction of single-family homes.

But this is only the beginning. Neotraditional zoning, including the proposed zoning code for my community, included "design codes" requiring peaked roofs, bay windows, full-width front porches, and certain other "cute" features. They also dictated that garages must be behind houses, not in front. Apparently flat roofs and prominent garages make people drive too much.

The new codes differed from the old in another important way: setbacks. Traditional codes require that homes and other buildings be built (typically) at least 20 feet apart and 30 feet from the street. In contrast, the new codes would require that homes be at *most* 20 feet from streets and that most commercial buildings be zero to ten feet from streets.

In sum, the planners got it wrong before, and now they propose to fix the urban blight they created. The solution they propose is a simple one: give them more power—power not just to prevent certain uses but to prescribe uses as well.

This will fix the problem?

"*The lawn . . . is a source of substantial pollution. . . . Excess chemicals . . . pollute water supplies, and pesticides contaminate food chains.*"

The American Lawn Is Bad for the Environment

F. Herbert Bormann

The lawn has been a feature of the typical American home for over a century and enjoys as much popularity today as ever. Originated in the more suitable, moist climate of England and introduced to America by early settlers, the lawn is ill fitted to the harsher conditions of the North American continent, according to F. Herbert Bormann. In the following viewpoint, Bormann, a professor emeritus at the School of Forestry and Environmental Studies at Yale University, traces the history of the American lawn and argues that efforts to adapt it to the nation's diverse countryside have resulted in ever-increasing hazards to the environment.

As you read, consider the following questions:

1. Who originated the lawn, according to Bormann, and why?
2. How, in Bormann's assessment, has the U.S. market economy impacted the lawn?
3. In what ways does the American lawn harm the environment, according to the author? What examples does he provide as lawn alternatives?

Excerpted from "Rethinking the American Lawn" by F. Herbert Bormann. This article originally appeared in the April 1994 issue and is reprinted by permission of the *World & I*, a publication of the Washington Times Corporation, ©1994.

The lawn, American as apple pie, is never far from sight, surrounding homes, lining country roads and interstate highways, and gracing parks and public spaces. In the United States some 25 million acres, the size of the state of Virginia, are covered with grass swards, 80 percent of which surround 58 million homes.

Americans love their lawns. They are gathering places for family, friends, and neighbors; a private arena for sports and relaxation. Grass feels good to the touch, cut grass freshens the smell of the air, and the whole brings an image of peace of mind. In passing through suburban neighborhoods, anywhere in the country, where one landscaped lawn follows another, we can vividly see the pride Americans take in their lawns.

But there is trouble in paradise! The American lawn has become the subject of criticism, the most serious of which is its implication as another factor in the deterioration of the local, regional, and global environment. To understand this turn of events, it is useful to examine some of the historical reasons why the lawn has reached its position of esteem in the American psyche, how the commercialization of the lawn has led to conflict with environmental goals, and what might be done to permit us to enjoy the many virtues of the lawn while reducing its impact on nature.

History of the Lawn

The lawn, as we know it, is not an old tradition: It was conceived by eighteenth-century English landscape gardeners. As English cities became increasingly polluted and disease-ridden, gardeners took advantage of the mild, moist climate of England and transformed the estates of the elite into great rolling landscapes of rocks, trees, and water all bound together by the moisture-loving lawn. All marks of human activity were erased to produce peaceful pictures of nature.

This vision of the English landscape was brought to America and incorporated into the estates of wealthy Americans like Thomas Jefferson who designed Monticello and the University of Virginia around the lawn, and George Washington, whose Mount Vernon estate featured a great tree-lined sward of grass sweeping down to the Potomac River. Lawns were not uncommon in late eighteenth-century America, but with few exceptions they were kept to a minimum. Grass was cut by hand scythes or grazing animals.

The advent of the lawn as the common people's art form was made possible by technology: In 1830, the Englishman Edwin Budding invented the lawnmower. Within decades, the well-manicured lawn was within easy reach of citizens of modest means. Whether they had an acre or a tiny patch of land, they could have a well-trimmed lawn as a centerpiece for their "estates."

In post–Civil War America, the lawn became the symbol of rapidly expanding suburbs. The curvilinear layout of American residential streets, with houses set well back from the road behind front lawns with informal plantings of trees and shrubs, became a uniquely American residential characteristic. The front lawn unified the whole neighborhood, giving a sense of community parkland.

Spikey Bushes and Funny Trees

[In Australia,] we don't have that crazy development ethic you Americans have, although we have damaged much of our land by trying to turn it into England. The British came out, and they didn't like these spikey bushes, funny animals hopping along and funny trees. So they chopped them down and tried to create these gorgeous green fields, elm trees, rose bushes and gravel paths. And we killed the aborigines.

Helen Caldicott, quoted by Will Nixon, *E Magazine*, September/October 1992.

Despite its origin in the mild, moist climate of England, the lawn has spread to every corner of North America, yet North America is a continent of many harsh climates and enormous diversity of vegetation and soils. Climates range from the extreme winter cold of Canada and northern United States to the intensely hot and droughty summers of the South, from the extremely wet Northwest Coast, to the wet East Coast, to the extremely dry Southwest. Nevertheless, the same continuous sward of green front lawn joins block after block in towns, suburbs, and cities across the land, whether wet, dry, hot, or cold.

Just how could the mild-mannered lawn conquer these hostile environments? Budding's invention was paralleled by an explosion in agricultural knowledge and technology. We learned about plant nutrition, about fertilizers and how to produce them, about plant disease and insect depredations of plants and how to combat them, about water requirements and how to meet them with irrigation, and how to breed plants better adapted to particular environments. And we invented dozens of labor-saving machines to perform all sorts of lawn chores.

As our country filled out, it became possible, using agricultural technology, to grow lawns in all sorts of environments that are naturally hostile to the prolonged growth of lawn grasses.

In a market economy like ours, the lawn was quickly realized to be a potential source of considerable profit. Through entrepreneurialism, technology became relatively inexpensive and available to homeowners through retail outlets. Chemical companies responded to the needs of small-scale purchasers of fertilizers and pesticides. Agricultural machinery companies began to produce a stream of lawn-care machines, and seed companies developed varieties for the home market. Sod companies made possible instant lawns, and lawn-care companies arose to relieve the homeowner of day-to-day responsibility for lawn care.

And so the lawn was progressively commercialized, and through entrepreneurialism it overcame local and regional environmental barriers. But the spread of the lawn was possible also because it was preferred and liked by a mass audience conditioned by a long cultural history.

Today there is an estimated $25 billion-a-year turf grass industry that over recent decades has, through advertisements, articles, speeches, and lobbying, fashioned a highly restricted definition of the lawn that is widely accepted by the general public. That lawn is the Industrial Lawn, which is based on these four principles of design and management:

1. It is composed of grass species only.
2. It is free of weeds and other pests.
3. Insofar as possible it is continuously green.
4. It is regularly mowed to a low, even height.

The Industrial Lawn is not attuned to the peculiarities of place: It ignores microclimates and species diversity and substitutes fossil-powered technology for solar-powered natural processes. The maintenance of the Industrial Lawn is dependent on the expenditure of considerable money, and its pursuit promotes the sales of fertilizers, machinery, pesticides, irrigation equipment, seeds, sod, and lawn-care services. But best of all in the eyes of the lawn-care industry, it has the virtue of never being completely attainable. There is always some new and necessary bit of machinery, some new finding about fertilizers or pesticides, or some new variety of grass required to keep up with the neighbors. All of which means money in the bank for the industry.

Challenging the Lawn

The monolithic concept of the Industrial Lawn is coming under increasing scrutiny by Americans who have begun to question the purpose of the lawn and its social, economic, and environmental costs.

Many people dislike the burden of constant lawn care. Time is often in short supply, and lawn care can be a demanding taskmaster. The routine of mowing, especially when irrigation and fertilization promote fast growth, can be demanding and a poor substitute for other activities. Many seek relief by hiring lawn-care companies, but others seek relief by developing new lawn-care standards and strategies that require less time and energy.

106

For other lawn owners the eighteenth-century vision of the lawn as nature has lost its strength. They seek new connections with nature, more personal, more local, and more dynamic. Some have become fascinated with a process called plant succession, where when mowing ceases, naturally occurring processes convert their lawns into meadows. Others prefer to use their land for gardens or plant displays other than lawn grass. Still others use their former lawns to study the behavior of wildflowers and insects.

However, the major challenge to the Industrial Lawn comes from the rise in environmental consciousness.

Environmental Impacts

About 30 years ago, signs of a steady decline in environmental quality became visible in many areas of our country: Polluted streams and rivers, smog-shrouded cities, and urban decay were everywhere. Scientists began to document food chains contaminated with pesticides and air masses with an extraordinary variety of wastes from our huge industrial base and our vast fleets of cars. Today, what seemed a patchwork of environmental problems has coalesced into global phenomena: global warming caused by the accumulation of greenhouse gases in the atmosphere; pesticides found in Antarctic penguins; large regions of the earth affected by acid rain and other air pollutants that cause extensive damage to forests, streams, and lakes; chlorofluorocarbon pollutants that thin the stratospheric ozone layer, with a resultant increase in biologically destructive ultraviolet light at the earth's surface that holds severe consequences for human health; and extensive pollution of substantial portions of the oceans.

All of this has revealed to us a global environmental deficit, the unanticipated consequences of humanity's alteration of the earth's atmosphere, water, soil, flora, fauna, and ecological systems. Backed by a mass of scientific evidence, we now realize that human activities can disrupt the very life systems on which we all depend. So widespread is this knowledge that a worldwide conference was held in Rio De Janeiro in 1992 to map out strategies to reverse these trends.

Pollution, Water Consumption, and Biological Poverty

How does the American lawn contribute to this scenario? The lawn, but particularly the Industrial Lawn, is a source of substantial pollution. The numbers are amazing. The Environmental Protection Agency (EPA) estimated that in 1984 more fertilizers were applied to American lawns than the entire country of India applied to all food crops. Most Industrial Lawns receive between 3 and 20 pounds of fertilizers and between 5 and 10 pounds of pesticides per year. Homeowners use up to 10 times more chemical pesticides per acre than do farmers. Excess chemicals are washed off lawns and pollute water supplies, and pesticides contaminate food chains. Lawn chemicals have become increasingly implicated in human health problems.

American lawn management uses about 1 percent of the gasoline we use in cars, but this amount is more significant than the percentage indicates. Much of it is used in highly polluting two-cycle engines, where the one hour it takes to mow a lawn produces air pollution equivalent to driving 350 miles. When all lawns are considered, it is easy to see that lawn care can be a major contributor to regional air pollution.

Water is one of our most precious natural resources and in many sections of the country is in short supply. In the Southwest, where rainfall is sparse, maintaining a lawn requires virtually continuous watering. Indeed, in the West, lawn watering can account for up to 60 percent of urban water use. Lawn watering is not limited to arid regions, however. In the more humid East, where water shortages are appearing more frequently, up to 30 percent of the water used in urban areas is for lawns. Many of these Industrial Lawns have been designed to be thirsty.

Nationally, we have a serious landfill problem. Existing landfills are filling up, and new sites are harder and harder to find. We generate 160 million tons of municipal solid waste annually, most of which ends in landfills. Yard waste is the second-largest component, and three-quarters of yard waste is grass clippings from our lawns, many of which are designed to produce maximum amounts of grass.

Finally, there is worldwide concern about a biological holocaust—the mass extinction of species, caused by human activities such as habitat destruction and pollution. One estimate projects that within 50 years, one-quarter of all species alive on the earth today will have become extinct. This is a matter of extraordinary concern to biologists everywhere.

How is the lawn implicated in mass extinctions? By design, the Industrial Lawn excludes all plant species but grasses, but because of its simple structure it also excludes many species of birds and other animals. New housing developments frequently carry out wholesale destruction of complex habitats relatively rich in species and replace them with Industrial Lawns extremely poor in species, making localities biologically impoverished. . . .

Understanding where the lawn's popularity comes from, how the lawn fits into the global environment, and finally, what changes we can make to alter its effects gives each of us the power to improve our piece of nature. We need not cease to love the lawn. But by understanding how it works we can adapt it to our time.

"Most lawn care pesticides are 'general use'
products—the federal Environmental Protection
Agency (EPA) considers them safe for use."

Pesticides Used for American Lawns Are Not Harmful to the Environment

Leonard T. Flynn

Growing awareness of the human impact on the environment
has led to greater sensitivity regarding the use of chemical products for maintaining American lawns. Yet according to regulatory and scientific consultant Leonard T. Flynn, concerns over
pesticides, which are monitored by scientists regularly for environmental safety, are largely unfounded. In the following viewpoint, he explains that only a small number of these products
are deemed toxic to any plants or animals other than the destructive species towards which they are targeted.

As you read, consider the following questions:

1. What reasons does Flynn provide for deeming "restricted
   use" substances to be of low risk to the environment?
2. What does the author identify as the key issue in the sign
   posting debate?
3. What does Flynn foresee as problematic about the sign
   posting laws?

Leonard T. Flynn, "Lawn Care Chemicals," in *Issues in the Environment*, 1992. Reprinted
with permission of the American Council on Science and Health, New York, NY.

Chemicals used on lawns are pesticides and fertilizers. Lawn care pesticides are substances used by man to control pests, primarily destructive insects, fungi and weeds, while fertilizers are plant nutrients to help the grass—and any other plants present—to grow.

Most lawn care pesticides are "general use" products—the federal Environmental Protection Agency (EPA) considers them safe for use by anyone who follows label directions. However, some types of lawn products are "restricted use" substances. They are only sold to and used by "certified applicators," persons who satisfy EPA and state training requirements. For "restricted use" pesticides the label instructions alone are not considered adequate to assure safe and proper use.

Fertilizers are not pesticides and they generally are not considered toxic substances. This appraisal is probably accurate, but users of fertilizers should still use care in handling them, for instance, to avoid eye exposure and to prevent access by children. Clearly, fertilizers are not entirely risk-free substances. Nevertheless, recent concerns have been raised about pesticides, not fertilizers, so the remainder of this discussion will focus on these materials.

The Low Risk of Pesticides

Unlike most other substances used by man, pesticides are designed to kill pests and must be toxic to the pests in order to work. A few lawn care pesticides are toxic to nearly all animal and plant life, such as fumigants used to eliminate nematodes [parasitic worms] prior to replanting grass in an infested area. Most lawn care pesticides, however, are relatively specific because their primary toxic effects are directed only to target species. For example, under normal conditions of use, phenoxy herbicides kill most broadleaf plants but do not adversely affect grasses, insects, or rodents.

Many lawn care pesticides have been used for decades on turf and for agricultural purposes, so experience with human exposure has been substantial. Nevertheless, scientists continue to study the toxicity of pesticides and other commercial materials to attempt to determine whether there are any unsuspected chronic effects.

Lawn care professionals handle undiluted pesticides while they mix the solutions for lawn applications, so their potential exposure to the chemicals is much greater than that of the homeowners who hire them or their customer's neighbors. It would seem hard to disguise significant numbers of chronic health effects for such workers from health insurance investigators, occupational safety and health professionals, industrial hygienists, and medical professionals in occupational health—not to mention government regulators. Despite this comforting lack of adverse information, effects possibly related to lawn care pesticides have recently appeared in the news.

Risk Reduction Through Lawn Posting Laws?

Despite the rarity of allergic effects from pesticides, several local governments have adopted various requirements for notifying the public about pesticide use. From 1986 to 1988, sign posting ordinances were adopted in Wauconda, Illinois, and in Montgomery and Prince George's Counties in Maryland. Both ordinances were challenged successfully in court by industry groups. The courts decides that under FIFRA (Federal Insecticide, Fungicide and Rodenticide Act), state or federal statutes specifically preempted the local jurisdictions from enacting pesticide regulations.

Whatever legal arguments ultimately prevail (both decisions were appealed by the local authorities), the fundamental issue is the public's "right to know" about potential hazards and the ability of sensitive individuals to, in effect, give "informed consent" to possible allergic reactions through their awareness that pesticides will be applied nearby. For these purposes the sign posting ordinances are inferior to a program of direct notification of sensitive persons who request that they be notified of imminent pesticide treatments.

Pesticides for Pests

Pesticides, by definition, kill pests. Insecticides kill insects, fungicides kill fungi and herbicides kill weeds. It is a serious error, however, to assume that because pesticides kill pests they are necessarily a threat to wildlife or to humans.

Andrea Golaine Case, *Washington Times*, June 20, 1995.

Direct notification is a standard procedure used to notify apiarists (beekeepers) of impending pesticide applications which might endanger their hives. For example, New Jersey pesticide regulations require at least 36 hours advance notification to each apiarist within one-half mile of the application site for a pesticide having a label indicating it is toxic to bees. Apiarists must register with the New Jersey Department of Environmental Protection (NJDEP) prior to March 1 of each year and the NJDEP may charge five dollars to offset its registry cost. Many other states have similar programs.

Where in effect, notification requirements provide sensitive individuals with the notice and opportunity to exercise informed consent to potential exposure. In contrast, lawn posting does not provide specific information to sensitive individuals and invites spurious complaints of illness from passers-by. The industry concern about such incidents is not without justification; "mass psychogenic [originating in the mind, or from mental or emotional conflict] illness" has been documented and hysterical reactions to sign postings for application of lawn care chemicals are not inconceivable.

The sign posting/notification issue really raises an ethical question beyond the scope of strict scientific evaluation: how can society deal with super-sensitive individuals? If your neighbor is deathly allergic to paint fumes, may you paint your house? May you allow ragweed to grow on your property and provoke serious asthma attacks next door? Must you get rid of your pet cat if the lady in the apartment across the hall is allergic? Are people compelled to gain "implied consent" and incur significant expense and inconvenience to protect their less hardy neighbors? Is the sensitive person solely responsible for him/herself and parents responsible for their sensitive children?

None of these questions is simple to answer and posting laws will do little to resolve them. In any case, the risk to anyone from lawn care chemicals is minute compared to other common allergens and irritants which are untouched by the posting laws.

"*Because these urban migrants don't understand rural life, they may help destroy the very human and aesthetic values they desire.*"

Environmentalism Threatens the Traditional Western Lifestyle

Alston Chase

According to former university philosophy instructor Alston Chase, environmentalists—not ranchers—pose the greatest threat to the American West. In the following viewpoint, Chase argues that these former city dwellers, newly arrived from overcrowded, polluted, and crime-ridden urban enclaves, are pressuring local and federal governments to deny grazing and farming rights on land that local families have worked for generations. He contends that as a result of these efforts an entire way of life—the icon of American tradition and values, the family farm—is in danger of extinction.

As you read, consider the following questions:

1. What reasons does Chase give for the environmentalists' efforts to conserve rangelands? Who are the environmentalists and how does he describe them?
2. What does Chase envision as the events that will lead to the death of the "small-town West"?
3. To what group does the author compare Western ranchers, and why?

Excerpted from "Turning Back the Clock" by Alston Chase, *Range* magazine, Winter 1994. Reprinted by permission of Alston Chase and Creators Syndicate.

In the movie *Shane*, Alan Ladd plays a mysterious gunslinger unable to escape his past. This motif of the man who brings trouble with him recurs in Western fiction, and in real life as well. Early in this century, people flocked to Phoenix seeking relief from hay fever. Unfortunately, many of them brought potted plants. Soon the town's pollen count soared and Phoenix became one of the sneezingest cities in the West.

Ranching the View

Now the theme is repeated again. Celebrating "boom time in the Rockies," *Time* magazine reports that this mountainous area is growing faster than any other. The latter-day settlers, *Time* says, are "modern cowboys" who "yearn for a simpler, booted front-porch way of life" and come searching for safety, family values, and solitude.

But these new immigrants are bringing the very miseries they seek to escape. Denver already has smog and street gangs, and because these urban migrants don't understand rural life, they may help destroy the very human and aesthetic values they desire.

When city folk flee to the mountains, they cram into celebrity ghettos like Santa Fe and Jackson, Wyoming, where they jog, bicycle, fly fish, backpack, and climb mountains. They buy big spreads so they can "ranch the view," but they can't tell clover from bunch grass.

A Culture in Crisis

Their ignorance does not prevent some from joining environmental groups that campaign to remove ranchers from the land. They ignore catastrophic overgrazing by wildlife in national parks because these playgrounds satisfy their atavistic yearnings for a time that never was, but they are offended by the sight of Herefords. So they seek to turn back the clock.

Hence, even as the *Time* article appeared, a small group of ranchers and filmmakers gathered in Steamboat Springs, Colorado, to baptize a new documentary entitled "Western Ranching: Culture in Crisis." The idea for the film came to its producer, Roger Brown, after he watched the 1991 Audubon special "The New Range Wars" on public television. This broadcast, which offered a sensationalist view of overgrazing, "was a vicious piece of propaganda under the guise of a documentary," he says. "We simply had to do something about it."

Appropriately, Brown's film is appearing just as Interior Secretary Bruce Babbitt prepares a coup de grâce against ranching. Dubbed a "grazing fee hike," Babbitt's proposal is in reality an omnibus administrative decree that will devastate the small-town West [Babbitt has since dropped this proposal and referred the grazing fee issue to Congress].

Ever since the late 1600s, farm families have cared for public lands that in many cases were intermingled with their own. They developed springs and irrigated pastures that fostered wildlife. And through their labors, they acquired lease and water rights, adding to the equity of their operations.

Ever More Arduous

Western ranchers, rugged and independent, are fixtures in American culture. Their livelihood, always arduous and subject to vagaries of weather and markets, has grown even more difficult of late, with beef prices plunging.

Equally disturbing to many ranchers are what they view as efforts to weaken their right to use millions of acres of publicly owned rangeland.

Christian Science Monitor, August 1, 1995.

The Babbitt plan will take these rights away. A lease can be terminated for many reasons, including minor legal infractions by the lessee, thus suggesting that someone might lose the lease his or her family had worked for a century by getting a traffic ticket.

And when the lease expires, the value of the ranch will plummet. The Internal Revenue Service (IRS) will collect fewer inheritance and income taxes; property values will decline, putting rural schools in jeopardy; local banks will be left holding insufficiently secured mortgage paper; and stores will go out of business. Towns will die, and the family farm—among the few institutions still nurturing people and the land—will disappear.

Dismantling the West

All this is done to please environmentalists, who embrace "biocentrism"; on their scale of values, humanity ranks somewhere near the primordial slime.

This notion requires depopulating the earth, one step at a time. It begins by evicting loggers, ranchers, and miners, then motorized recreationists, hunters, and eventually fishermen, until the "Buffalo Commons"—a proposed 139,000-square-mile sanctuary spreading over 10 Great Plains states—becomes reality. If that seems to be an exaggeration, consider this: In response to prompting from Colorado environmentalists, officials have already identified some grazing allotments in Montana as "unoccupied bison habitat."

After the Rockies are cleared of the threat of humanity, activists will move on to the Appalachians and the Sierra. Their work is never done.

This suits the federal government, which ever since it deep-sixed Native American tribes has made a habit of deconstructing viable cultures. Indeed, that is how things have changed since *Shane*. In the movie, settlers supposedly introduced civilization to the West. Today, many of the new immigrants want to dismantle it.

"Western ranchers have traditionally fed well at the trough of Federal beneficence. In their war against Washington, they are biting the hand that has fed them."

Environmentalism Does Not Threaten the Traditional Western Lifestyle

The New York Times

Who should be the steward of the land has long been hotly debated in the West, with local governments squaring off against Washington for control over land use and management. This viewpoint, which originally appeared as an editorial in the *New York Times*, argues that those waging the latest skirmish in this battle with the federal government are driven by big business and its profit motive. Meanwhile, according to the *Times*, the genuine desires of local residents, whom these groups claim to represent, go unheeded.

As you read, consider the following questions:

1. What was the original "Sagebrush Rebellion," according to the *Times*, and how does it differ from the current "war in the West"?
2. Why do the *Times* editors fear transferring control of endangered species protection to local governments?

Editorial, "The Endangered West," *New York Times*, E-14, June 18, 1995. Copyright ©1995 by The New York Times Company. Reprinted by permission.

A sample of recent bulletins from the Old West: Montana rewrites some of the country's strongest water pollution laws as a favor to the mining industry. Idaho lawmakers award potential polluters a major voice in setting clean water standards. Utah's Governor rebuffs the stated wishes of Utah's citizens to set aside 5.7 million acres of state land as protected wilderness. Washington State's Legislature passes the nation's most far-reaching "takings" law, weakening essential land-use controls. Wyoming's Legislature authorizes a bounty on wolves—recently re-introduced into Yellowstone National Park and protected under the Federal Endangered Species Act.

Clearly, the United States Congress is not the only place where laws protecting the environment are under siege. Throughout the West, particularly in the Rocky Mountains, state legislators and governors, egged on by commercial interests and by small but noisy groups of property-rights advocates, are engaged in full-scale mutiny against Federal and state regulations meant to protect what is left of America's natural resources.

Another Sagebrush Rebellion

What we are seeing is an updated but more ominous version of the Sagebrush Rebellion of the early Reagan years. That revolt was dominated by ranching interests protesting Federal regulation of public lands. The present explosion embraces not only those familiar despoilers but mining companies, timber barons, developers, big commercial farmers and virtually anyone else who stands to profit from relaxation of environmental controls.

The war in the West and the war in Congress on basic environmental protections have much in common. First, both are being driven and in some cases underwritten by big business. Second, both are being waged to save the "little guy" from Federal tyranny. Third, this alleged little guy is nowhere to be found when the time comes to draft crippling legislation. Indeed, his wishes have been largely ignored. Poll after poll suggests that what ordinary citizens want is more environmental protection if it means a cleaner environment and a healthier society. But that is not what the 1995 Congress and its Western allies want to give them.

Montana and Idaho are particularly sad cases. Despite citizen complaints, and virtually unanimous editorial opposition, two bills whistled through the Montana Legislature that would in effect permit higher levels of toxic wastes to reach the state's streams and lakes. They were signed, with some reluctance, by the Governor. Mining lobbyists were conspicuous during the parliamentary maneuvering—including representatives from Crown Butte and its Canadian parent, Noranda Inc. These companies are working relentlessly for permission to build in geologically precarious terrain a gold mine that would leave a permanent reservoir of pollutants in the watershed of one of Montana's most important wilderness streams.

Indigenous Wildlife, Goodbye

Idaho's people—not to mention its endangered Snake River salmon—face a double threat. Under a new statute, acceptable water quality levels will be set by watershed advisory groups. These groups will be well stocked with large landowners and representatives from timber, mining, and agribusiness companies, who are virtually certain to write new and more permissive regulations. Meanwhile, back in Washington, an Idaho Republican, Dirk Kempthorne, is leading the Senate charge to cripple the Endangered Species Act, which provides what little protection the salmon have. If Senator Kempthorne succeeds in transferring protection of endangered species from Washington to Boise, it will be goodbye salmon, with grizzlies and wolves to follow.

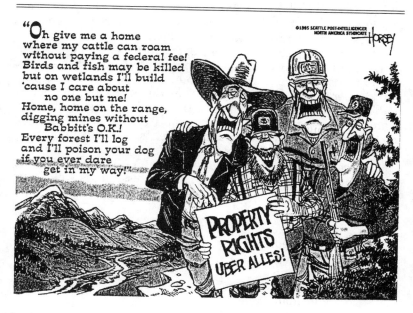

Reprinted with special permission of North America Syndicate.

There are, of course, honorable exceptions. In Colorado, for example, ranchers, environmentalists and state officials were able to agree on less destructive grazing practices—although it took a half-dozen or so exhausting visits from Interior Secretary Bruce Babbitt to get the agreement. But nearly everywhere one turns the anti-Washington ideologues seem to have the upper hand.

The most conspicuous example is Nevada, where officials in Nye County passed a series of ordinances claiming ownership of Federal lands and then set about physically intimidating employees from the Forest Service and the Bureau of Land Management. The Justice Department has now sued to reaffirm Federal jurisdiction, but Nye County's rebels have inspired imitators: More than 70 rural Western counties have passed or proposed laws to "take back" the public lands.

Biting the Hand That Feeds

Lost in all the rhetoric about individualism and states' rights is one basic legal fact: At no time have the Western public lands belonged to the states. They were acquired by treaty, conquest or purchase by the Federal Government acting on behalf of all the citizens of the United States. Lost, too, is a colossal irony. Western ranchers have traditionally fed well at the trough of Federal beneficence. In their war against Washington, they are biting the hand that has fed them lavish subsidies and protected them against the disasters of nature and the vagaries of the marketplace.

But all of this escapes the Sons-of-Sagebrushers. The fact that there might be an overriding national interest in preserving the public lands and forests from exploitation is not something that quickly pops to their minds. Nor does this fact seem to register with the newer breed of rebels in the statehouses and state legislatures who would nullify more than two decades of struggle to clean America's waterways, preserve its wetlands and otherwise protect its dwindling natural heritage.

There can be no satisfaction in any of this—except perhaps to the enemies of the environment in a Congress that is well on its way to abandoning any pretense to national stewardship.

Periodical Bibliography

The following articles have been selected to supplement the diverse views presented in this chapter. Addresses are provided for periodicals not indexed in the *Readers' Guide to Periodical Literature*, the *Alternative Press Index*, or the *Social Sciences Index*.

William Chaloupka
"Cynical Nature: Politics and Culture After the Demise of the Natural," *Alternatives*, Spring 1993.

Richard Hugus
"Destroying Us in Order to Save Us," *Z Magazine*, September 1993.

Maria Elena Hurtado
"Changing Consumerism," *Our Planet*, vol. 6, no. 2, 1994. Available from PO Box 30552, Nairobi, Kenya.

Charles Komanoff
"Undoing Automobile Dependence," *Workbook*, Summer 1994.

Gary Lee
"An Environmental History of America," *Washington Post National Weekly Edition*, December 12–18, 1994. Available from Reprints, 1150 15th St. NW, Washington, DC 20071.

James Malone
"Environmental Degradation and Social Injustice," *Origins*, March 18, 1993. Available from 3211 Fourth St. NE, Washington, DC 20017-1100.

Randal O'Toole
"The Greening of Liberty," *Liberty*, May 1995. Available from PO Box 1181, Port Townsend, WA 98368.

Mary Beth Regan and Peter Burrows
"Uncle Sam Goes on an Eco-Trip," *Business Week*, June 28, 1993.

James Thornton
"Technology and Freedom," *New American*, July 12, 1993. Available from 770 Westhill Blvd., Appleton, WI 54914.

Anne-Marie Willis
"Will Your Fingers Do the Shopping?" *World Monitor*, May 1993. Available from CSPS, 1 Norway St., Boston, MA 02115.

John Young
"The New Materialism: A Matter of Policy," *World Watch*, September/October 1994. Available from PO Box 6991, Syracuse, NY 13217-9942.

Is Ecological Conservation Bad for the Economy?

Chapter Preface

At Louisiana-Pacific Corporation, a pulp mill in Samoa, California, hardly anyone is buying. After being sued by environmental groups for dumping chlorine-contaminated wastewater into the ocean, the company began producing totally chlorine-free (TCF) pulp, which apparently is less popular with the mill's industrial customers than it is with environmentalists. Louisiana-Pacific says that its attempts to sell TCF pulp have, in fact, resulted in such steep losses that at one point the company halted its production.

The greening of American industry is still in its infancy, according to some. "This market is evolving right now," says Peter Sweeney of Cross Pointe Paper, whose company is also developing TCF paper products. Going "green" means voluntarily undertaking any efforts that exceed energy efficiency and waste regulations in manufacturing operations, as well as shifting to renewable resources for product and packaging materials. Companies that employ such practices often appeal to the environmental conscience of consumers by informing them about these acts of goodwill through their advertising.

While the Federal Trade Commission enforces guidelines regarding accuracy and honesty in promoting environmental good deeds, some people remain skeptical about the truth behind claims made by "green" companies. David Rothbard and Craig Rucker call green marketing a "phony" ploy used to sell "often higher priced eco-items that solve nonexistent problems."

Yet, as Louisiana-Pacific and some environmental writers observe, the problem is precisely the opposite: Consumers are *not* spending their dollars on green products—at least not enough to make environmentally motivated business decisions pay. Some people attribute this failure to Americans' short attention span for social issues. "Green consumerism was one of those well-intended passing fancies," writes the *Green Business Letter*'s editor, Joel Makower. He also criticizes companies that have failed to use their marketing and advertising resources to inform consumers about the environmental benefits of purchasing green products.

Others insist that it is too soon to call green business a failure. "There's a market need and we have to identify how large that is," maintains Sweeney. In response to consumer demand, Cross Pointe has also added a line of products made of recycled materials. Environmental lawyers J. Stephen Shi and Jane M. Kane agree. "These efforts are important in the face of tightening markets and stiff competition," they contend. The sometimes precarious relationship between a healthy environment and a vibrant economy is examined throughout this chapter.

"The Forest Service has now banned the cutting of trees larger than 30 inches in diameter . . . and restricted the cutting of trees over 24 inches . . . eliminating thousands of jobs."

Efforts to Protect Forests Endanger Loggers' Livelihood

James Owen Rice

Environmental restrictions designed to protect spotted owl forest habitats in the Pacific Northwest have slowed the logging industry to a near standstill, contends environmental writer James Owen Rice. Yet numerous sightings of the spotted owl in previously cut areas of the forest prove that logging and the spotted owl can—and do—coexist, according to the researchers he interviews in the following viewpoint. Even so, Rice argues, the Forest Service continues to enforce bans on logging that are devastating the industry and the many local communities that depend on it for their subsistence.

As you read, consider the following questions:

1. How much has employment dropped since the legal battle over spotted owl preservation?
2. Why does Ben Stout dispute Al Gore's claim that logging jobs would have been lost regardless of logging bans?
3. What does Rice present as evidence that the spotted owl does not require preservation of old growth forests to survive?

James Owen Rice, "Environmental Extremism Destroying Northwestern Jobs," *Human Events*, April 17, 1993. Reprinted with permission.

Although many Americans were probably aware of the Clinton Forest Conference held April 2, 1993, to resolve the timber wars of the Pacific Northwest, special attention was paid by citizens of the remote northern California town of Happy Camp.

Located in the rugged Klamath River valley, and surrounded by the lofty evergreen forests of Siskiyou County, this hamlet's workforce is 35 percent Native American, many of them Yuroks, Karuks and Hoopas, according to former Karuk tribal council member Jim Waddell.

Looming Uncertainty

Like many of the town's 1,100 residents, Waddell is not a happy camper. A timber manager for the local Stone Forest Products mill, he has seen local employment drop 60 percent and family members scatter as spotted owl lawsuits slashed the timber harvest on four nearby National Forests by 85 percent. The mill has eliminated one of two shifts and [as of April 1993] has enough logs to run for a few months. After that, uncertainty looms.

"We're surrounded by almost 1.5 million acres of virgin forest, yet the government, which violated the original Indian Treaty by never paying us for the land, is starving us out," says Waddell. "I have recommended to Tribal Council members that they pursue this. If the government isn't going to manage the land sensibly, then maybe we should repossess it and manage it ourselves."

At Happy Camp High School, where one-third of the student body is classified as American Indian, principal Jay Clark worries less about the declining enrollment than he does about so many families broken up by breadwinners forced to relocate.

Los Angeles, 12 hours away, is a common destination for fathers seeking work. "I know of one case," he says, "where the mother is working in Valdez, Alaska, while the father, who can't find work, remains behind with the family."

"We have seen an increase in juvenile crime, spousal abuse and divorce," says Siskiyou County Superintendent of Schools Frank Tallerico, whose office runs a support service for students impacted by closures at the area's sawmills. His office also runs a school for incarcerated juveniles. "We've never had a full house at our juvenile hall before, but in October of 1991 we did. And we are a microcosm of what is happening all over the Pacific Northwest."

Native Americans play a major role in the western timber industry, from the federated tribes of Colville, Washington, and Warm Springs, Oregon, to the Piutes and Monos of the southern Sierra Nevada along with many others. Since by most social measures (including premature mortality and poverty rate) they are the most disadvantaged minority in America, divesting the rural economy of the only manufacturing jobs available will have serious social consequences.

The Forest Service's 1990 figures show that the timber sale program is a mainstay of the rural economies in all three states. That year, the program generated 62,000 local jobs while paying the states $254 million for local schools and roads (in lieu of property taxes). In addition, it yielded a profit to the federal government of $404 million and raised $347 million in federal income taxes.

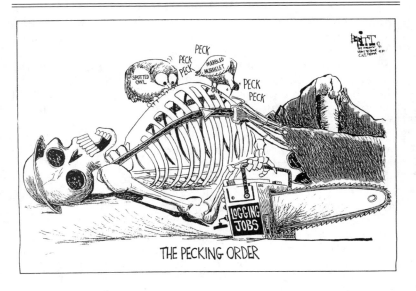

©Britt/Copley News Service. Reprinted with permission.

Although it is often alleged that the Forest Service actually loses money on its western timber sale program, and thus subsidizes loggers, its accounting methods have been systematically refined over recent years to conform to guidelines recommended to it by the General Accounting Office, including those recommended in a 1989 audit. Forest Service critics can't say as much. Nor can supporters of President Clinton's new $16.3-billion stimulus package claim that it will return a profit.

"Pure Hokum"

The mainstream press asserts that the current rending of the rural social fabric is inevitable because loggers are little more than "buffalo hunters" who are wiping out the last of the old growth and the spotted owl along with it. Vice President Albert Gore reflected conventional press wisdom when he wrote in *Earth in the Balance* that the logging jobs in the Northwest "would have been lost anyway as soon as the remaining 10 percent of the forest was cut."

Foresters dispute that.

"Pure hokum" is how Ben Stout, retired dean of forestry at the University of Montana, characterizes that claim. That's because 4.5 million acres (equal in area to Connecticut and Rhode Island) of old growth were already protected in Oregon and Washington before the owl controversy.

"No one is going to touch those old trees and I'd put my body in front of a bulldozer to protect them," he says. "Beyond that, Al Gore and others ignore the fact that trees grow, and we are already cutting younger forests."

New forest plans and spotted owl set-asides, he and others maintain, have made the orderly transition to a young growth economy impossible by ballooning the reserves to 19 million acres, an area the size of Vermont, New Hampshire and Massachusetts.

In California, before the owl debate there were two million acres of parks and two-thirds of the 20-million-acre National Forest lands were protected, including four million acres of wilderness. While most of that is not suitable owl habitat, millions of acres are.

Spotted Owls in Abundance

Although the media continue to report that spotted owls depend for survival on old growth forest, an unreported consensus among owl experts holds that forest structure (including a multi-layer canopy, dead and down trees, and nesting cavities) is more important for owls than tree age. Thus, previously managed forests that have been logged in such a way as to preserve these features usually have just as many owls as old growth.

On California's North Coast, zoologist Lowell Diller has banded over 500 northern spotted owls on the Simpson Timber Company's 380,000 acres of redwood forest, less than 2 percent of which is old growth: "If we have created spotted owl habitat by accident, certainly we can do so by design," he says.

"It's clear that the distribution of spotted owls is not identical with the distribution of old growth forests," says former wildlife biology professor and recognized owl expert Larry Irwin. His extensive studies in Washington's eastern Cascades have revealed that more than 40 per cent of owl nest sites there are located in previously managed stands.

Even biologists for the U.S. Fish and Wildlife Service (which declared the northern spotted owl threatened) now concede that owls and logging are compatible. Senior staff biologist Phil Dietrich says that the presence of spotted owls on private lands in California "leads us to believe that it is possible to manage for timber and maintain owls."

The most recent report on the California spotted owl (which is the same species, *Strix occidentalis*) authored by an interagency committee chaired by Forest Service biologist Jared Verner found that despite 100 years of logging in its Sierra Nevada habitat, California spotted owls "continue to be widely distributed throughout most of the conifer zone. Indeed, spotted owls may be more abundant in some areas of the Sierra than they were 100 years ago."

Locked Up

Despite those encouraging findings, the Forest Service has now banned the cutting of trees larger than 30 inches in diameter anywhere in the Sierra, and restricted the cutting of trees over 24 inches, measures that are eliminating thousands of jobs throughout the region, in addition to those lost in the Pacific Northwest.

Oregon State University Professor of Forest Ecology Mike Newton, who is conducting an Interior Department–funded experiment to maintain the preferred structure in producing forests, sees the issue this way.

"Right now the Forest Service in Oregon and Washington is trying to decide whether to set aside one-third of the entire volume of timber in the United States for one species. Instead, we need to understand what is crucial for owls and other wildlife and create that and perpetuate it. We can't do that if the forest is locked up."

While that commonsense approach holds little appeal for some members of the President's decidedly green Cabinet, Clinton claimed at the Forest Conference that he was interested in a real compromise and ordered his Cabinet to come up with a plan in 60 days.

Back in Happy Camp, former wilderness guide Jim Waddell is working to develop a career as a professional photographer. But for most, the only hope is a Republican rescue.

Such a rescue, which might involve supporting some type of Clinton compromise against the environmental left of the Democratic Party, could provide another opportunity for the GOP to unite, this time in defense of moderate environmentalism and blue-collar families. But any political calculation will have to factor in a green press. Says Waddell, "God help you if the media decide your job is environmentally incorrect—the truth won't."

"Present over-cutting and rapid rate of harvest guarantees that . . . there will be a long period between harvests . . . without jobs for the workers."

Protecting Forests Also Protects Logging Jobs

Virginia Warner Brodine

The relentless drive to maximize short-term profits—not environmentalism—is destroying jobs in the logging industry, according to activist and freelance writer Virginia Warner Brodine. In the following viewpoint, she argues that the profit motive prods logging companies towards increased mechanization, a cheaper foreign workforce, and conversion of forest lands to other uses—all of which mean fewer jobs for American loggers. Further, she contends, depleting forests through clear-cutting and other nonsustainable practices destroys future prospects for the logging industry and the loggers it employs. Forests must be managed to protect both forests and logging jobs, she concludes.

As you read, consider the following questions:

1. In what ways is a tree farm like a forest, according to the author?
2. What does Brodine propose as an alternative to harvesting old growth forests?
3. Why is the spotted owl considered so important to forest preservation, according to Brodine?

Virginia Warner Brodine, "Timber: Saving the Environment and Jobs, Too," *People's Weekly World*, April 17, 1993. Reprinted with permission.

At the Timber Summit in Portland, Oregon, on April 2, 1993, President Clinton said he cannot "legislate against the laws of change."

Timber workers are asking: "Do the laws of change require that jobs disappear?"

As long as the engine of change is driven by the big timber companies—Weyerhaeuser, Boise Cascade, Plum Creek, Potlatch and others—the answer is "yes!" The capitalist law, "profit is sacred," will continue to be followed, and the natural laws that govern the growth and renewal of forests will be ignored.

If Clinton permits all or some of the old-growth cutting the companies want, there will be a few more jobs now but the long slide down the clear-cut hillside that has been destroying both trees and jobs will go on.

The timber companies like to make environmentalists the scapegoats but the corporate engine of change began reducing jobs long ago, when the great forests of the northwest, from northern California into Canada, were cut as if they had no end.

Profits vs. Jobs

Had the privately owned timber land been managed in such a way as to continue to renew what should be a renewable resource, there would be no battle now over national forests. Forests have been clear-cut and turned into tree farms. A tree farm is not a forest. It has rows of a single species, well adapted to another clear cut because they are all the same age. Undergrowth is kept down with herbicides, pests killed with pesticides, depleted soil is unable to continue producing one harvest after another.

Technical changes began in the woods with the introduction of the chain saw in the forties. Retired logger Gordon "Brick" Moir recalls losing his job in a cutting crew when the chain saw reduced the crew from 96 to 20. Mechanization continued through today's mobile yarders and other heavy equipment which are more efficient in getting out the trees, but destructive both of jobs and of the forest's regenerative capacity.

Each new generation of equipment has introduced more mechanization into the mills—and fewer jobs. A few years ago 116 jobs were lost when Weyerhaeuser closed its sawmill in Raymond, Washington. Union and community fought for modernizing the process instead of abandoning the town. The fight was won, the new, more highly mechanized mill was built. While 88 jobs were saved, 28 were lost.

Mill workers are also losing jobs because raw logs rather than milled lumber are shipped abroad. Environmentalists have joined with unions to halt this process. There are now some restrictions on such exports from national, and in some cases from state, land. But in the name of free trade, exports from private land remain unrestricted.

Some companies have moved logging operations south, where faster-growing timber can cut the long period between one harvest and another. The mills go with them. Or the logging may continue here but the milling be moved to a low-wage area. The North American Free Trade Agreement (NAFTA) could increase this trend.

"....ON THE BRIGHT SIDE, THE ENDANGERED SPECIES LIST IS DOWN TO ONE."

Reprinted with special permission of King Features Syndicate.

The influence of the big timber companies on the national government permitted much cutting in the national forests as well. U.S. Forest Service districts got high quotas of timber sales to meet. And the timber companies on the buying end have been, in effect, subsidized by prices lower than the cost to the government of forest management, road building, etc.

All these were profit-driven changes. None took the slightest account of workers or their communities.

The debate which reached the timber summit focuses on about 4.5 million acres of spotted owl habitat in the national forests which, without the protection of the Endangered Species Act, would be up for timber sale and harvest. The spotted owl is important because it is an indicator of the health of the forest ecosystem as a whole.

The courts have held up timber sales in national forests till the

Department of Agriculture's forest service comes up with a plan to protect the owl and other old-growth-dependent species.

A report by the Thomas Committee, the scientific group which first recommended protection of spotted owl habitat, now says that is not enough. Truly looking at the forest ecosystem, not simply at owl habitat, would require still more old-growth acres to be protected.

The forests are not all alike. For example, there are only about 75 million acres of temperate rain forest remaining on the globe and two-thirds of it is along the northwest coast of the United States and Canada.

Environmentalists cry out that now or never, we must save the last remnants of the ancient forest, while timber companies demand a continuation of old-growth cutting. Timber workers, seeing this as the only way to preserve what jobs are left, follow the companies.

Outside the summit, people from timber towns—loggers, mill workers, "independent" contractors and truck drivers, small business people dependent on timber wages, families of all of them—rallied on behalf of their jobs. There were rallies for protection of the northwest's diminished ancient forests, too, which drew many thousands. Those favoring protection included Native Americans and commercial fisherman, usually on opposite sides, who agree that salmon have diminished in part because their streams are dependent on the dwindling forests.

A Global Need

Readers tend to identify with the workers, but at the same time are likely to question corporate slogans, and to shudder at the sight on TV of clear-cut mountains.

As we try to sort out where we belong in this fight, we have to step back and look at all the forests, everywhere in the country, not just on the coastal side of the mountains in the Pacific Northwest. There is so little forest left that we have to be concerned with it all, privately as well as publicly owned.

The National Forests are the battleground partly because the huge, old trees have the best timber and therefore bring the highest profits. Furthermore, companies don't want any interference on their private forest holdings. The Endangered Species Act affects private forests, too, but timber companies have so far kept restrictions on private land less stringent than on national forest land.

Environmentalists rallied to this battleground because that's the only place there are any true forests left in their natural state. And that's where national ownership gives some hopes of putting on the brakes.

The state of second-growth forests has been almost forgotten in the focus on the so-called spotted owl controversy. Protection of what is left of our ancient forests is a national and even a global need. We need lumber and jobs, too, but these needs can be met from already cut-over lands in national and state forests and in privately owned timber lands. Now in second and third growth, these forests can provide jobs, lumber, protection of soil and water, protection of some forest plant and animal species, and continued scenic and recreation values if properly managed. Good management means more jobs in caring for the forest as it grows, more jobs by using more selective and less mechanized methods of harvest.

A Slower Pace

Trying for national and state policies governing this land is the only way to the kind of change that can answer "no" to the desperate question: will timber jobs continue to disappear?

Private property rights in forest lands must be challenged. So must corporate domination of the Department of Agriculture. The grab for immediate profits cannot be allowed to override the interests of the whole society. If timber companies made their harvest plans with long-term profits over the next 100 years or more in mind, truly sustainable forest practices would make sense to them.

Nature moves slowly. Trees cannot be harvested for 60 to 80 years, depending on the climate, and will produce more and better lumber if allowed to grow longer. But the higher the immediate profit, the greater the temptation to cut early, even to cut and run.

Take Burlington Northern, with its millions of acres of timber from the 1864 land grant. It set up a timber arm (Plum Creek Timber Co.). But also a real estate arm which can sell the timber land after the next cutting, if that looks like a better deal than waiting for the trees to grow again.

As the price of both land and timber has risen, small owners of timber land are tempted to sell even the smallest trees. They may never raise another crop. There are now big bucks in real estate.

As for the big owners like Plum Creek, their present overcutting and rapid rate of harvest guarantees that as long as their holdings remain timber land, there will be a long period between harvests without profits for the company and without jobs for the workers.

If a piece of land is sold or converted by the company itself to what they like to call a "higher and better use" there will be profits for the company, but—on that particular piece of land—never again jobs for the workers.

"*Those concerned with protecting the environment . . . should seek to expand capitalism . . . to the broadest possible range of environmental resources.*"

Private Industry Is Good for the Environment

Fred L. Smith Jr.

Fred L. Smith Jr. is president of the Competitive Enterprise Institute, a public interest group he founded in 1984 that promotes the principles of free enterprise and limited government. In the following viewpoint, Smith argues that the free market system is the best steward of the earth's natural resources. The ingenuity and ambition that fuel capitalism, according to Smith, also fuel continual improvements in the efficiency of production and distribution of goods and services. Greater efficiency, he maintains, means using fewer resources to provide more goods—which is essentially the definition of "sustainable development."

As you read, consider the following questions:

1. What definition of "sustainable development" does Smith recount, and how does he interpret it?
2. What examples does Smith provide to support his contention that private ownership and management are more successful than governments' efforts to preserve the environment?
3. How does the author respond to the argument that capitalism does not protect resources for future generations?

Excerpted from "The Market and Nature" by Fred L. Smith Jr., *Freeman*, September 1993. Courtesy of the *Freeman* and the Foundation for Economic Education.

Many environmentalists are dissatisfied with the environmental record of free economies. Capitalism, it is claimed, is a wasteful system, guilty of exploiting the finite resources of the Earth in a vain attempt to maintain a non-sustainable standard of living. Such charges, now raised under the banner of "sustainable development," are not new. Since Thomas Malthus made his dire predictions about the prospects for world hunger, the West has been continually warned that it is using resources too rapidly and will soon run out of something, if not everything. Nineteenth-century experts such as W.S. Jevons believed that world coal supplies would soon be exhausted and would have been amazed that over 200 years of reserves now exist. U.S. timber "experts" were convinced that North American forests would soon be a memory. They would similarly be shocked by the reforestation of eastern North America—reforestation that has resulted from market forces and not mandated government austerity.

The Disappearance of Everything

In recent decades, the computer-generated predictions of the Club of Rome enjoyed a brief popularity, arguing that everything would soon disappear. Fortunately, most now recognize that such computer simulations, and their static view of resource supply and demand, have no relation to reality. Nevertheless, these models are back, most notably in the book *Beyond the Limits* [by Donella H. Meadows], and enjoying their newly found attention. This theme of imminent resource exhaustion has become a chronic element in the annual Worldwatch publication, *State of the World*. (This book is, to my knowledge, the only gloom-and-doom book in history which advertises next year's edition.) Today, sustainable development theorists, from the World Bank's Herman Daly and the United Nations' Maurice Strong to Vice President Albert Gore and Canadian David Suzuki, seem certain that, at last, Malthus will be proven right. It was this environmental view that was on display at the United Nations' "Earth Summit" in Rio de Janeiro in 1992. This conference, vast in scope and mandate, was but the first step in the campaign to make the environment the central organizing principle of global institutions.

If such views are taken seriously, then the future will indeed be a very gloomy place, for if such disasters are in the immediate future, then drastic government action is necessary. . . .

The world does indeed face a challenge in protecting ecological values. Despite tremendous success in many areas, many environmental concerns remain. The plight of the African elephant, the air over Los Angeles, the hillsides of Nepal, the three million infant deaths from water-borne diseases throughout the world, and the ravaging of Brazilian rain forests all dramatize areas where problems persist, and innovative solutions are necessary.

Sustainable development theorists claim these problems result from "market failure": the inability of capitalism to address environmental concerns adequately. Free market proponents suggest that such problems are not the result of market forces, but rather of their absence. The market already plays a critical role in protecting those resources privately owned and for which political interference is minimal. In these instances there are truly sustainable practices. Therefore, those concerned with protecting the environment and ensuring human prosperity should seek to expand capitalism, through the extension of property rights, to the broadest possible range of environmental resources. Our objective should be to reduce political interference in both the human and the natural environments, not to expand it.

Private stewardship of environmental resources is a powerful means of ensuring sustainability. Only people can protect the environment. Politics *per se* does nothing. If political arrangements fail to encourage individuals to play a positive role, the arrangements can actually do more harm than good. There are tens of millions of species of plants and animals that merit survival. Can we imagine that the 150 or so governments on this planet—many of which do poorly with their human charges—will succeed in so massive a stewardship task? Yet there are in the world today over five billion people. Freed to engage in private stewardship, the challenge before them becomes surmountable.

Sustainable Development and Its Implications

The phrase *sustainable development* suggests a system of natural resource management that is capable of providing an equivalent, or expanding, output over time. As a concept, it is extremely vague, often little more than a platitude. Who, after all, favors non-sustainable development? The basic definition promoted by Gro Harlem Brundtland, former Prime Minister of Norway and a prominent player at the Earth Summit, is fairly vague as well: "[S]ustainable development is a notion of discipline. It means humanity must ensure that meeting present needs does not compromise the ability of future generations to meet their own needs."

In this sense, sustainability requires that as resources are consumed one of three things must occur: New resources must be discovered or developed; demands must be shifted to more plentiful resources; or, new knowledge must permit us to meet such needs from the smaller resource base. That is, as resources are depleted, they must be renewed. Many assume that the market is incapable of achieving this result. A tremendous historical record suggests exactly the opposite.

Indeed, to many environmental "experts," today's environmental problems reflect the failure of the market to consider ecological values. This market failure explanation is accepted by a panoply of political pundits of all ideological stripes, from Margaret Thatcher to Earth First! The case seems clear. Markets, after all, are shortsighted and concerned only with quick profits. Markets undervalue biodiversity and other ecological concerns not readily captured in the marketplace. Markets ignore effects generated outside of the market, so-called externalities, such as pollution. Since markets fail in these critical environmental areas, it is argued, political intervention is necessary. That intervention should be careful, thoughtful, even scientific, but the logic is clear: Those areas of the economy having environmental impacts must be politically controlled. Since, however, every economic decision has some environmental effect, the result is an effort to regulate the whole of human activity.

Thus, without any conscious decision being made, the world is moving decisively toward central planning for ecological rather than economic purposes. The Montreal Protocol on chlorofluorocarbons [an agreement by twenty-seven countries to reduce their 1988 chlorofluorocarbon emission levels by 50 percent by 1999], the international convention on climate change, the proposed convention on biodiversity, and the full range of concerns addressed at the U.N. Earth Summit—all are indicative of this rush to politicize the world's economies. That is unfortunate, for ecological central planning is unlikely to provide for a greener world.

Rethinking the Market Failure Paradigm

The primary problem with the market failure explanation is that it demands too much. In a world of pervasive externalities—that is, a world where all economic decisions have environmental effects—this analysis demands that all economic decisions be politically managed. The world is only now beginning to recognize the massive mistake entailed in economic central planning; yet, the "market failure" paradigm argues that we embark on an even more ambitious effort of ecological central planning. The disastrous road to serfdom can just as easily be paved with green bricks as with red ones.

Environmental policy today is pursued exactly as planned economies seek to produce wheat. A political agency is assigned the task. It develops detailed plans, issues directives, and the citizens comply. That process will produce *some* wheat just as environmental regulations produce some gains. However, neither system enlists the enthusiasm and the creative genius of the citizenry, and neither leads to prosperity. In fact, political management has been able to turn the cornucopia that was the

Horn of Africa into a barren, war-torn desert.

That markets "fail" does not mean that governments will "succeed." Governments, after all, are susceptible to special interest pleadings. A complex political process often provides fertile ground for economic and ideological groups to advance their agendas at the public expense. The U.S. tolerance of high sulfur coal and the massive subsidies for heavily polluting "alternative fuels" are evidence of this problem. Moreover, governments lack any means of acquiring the detailed information dispersed throughout the economy essential to efficiency and technological change.

More significantly, if market forces were the dominant cause of environmental problems, then the highly industrialized, capitalist countries should suffer from greater environmental problems than their centrally managed counterparts. This was once the conventional wisdom. The Soviet Union, it was argued, would have no pollution because the absence of private property, the profit motive, and individual self-interest would eliminate the motives for harming the environment. The opening of the Iron Curtain exploded this myth, as the most terrifying ecological horrors ever conceived were shown to be the Communist reality. The lack of property rights and profit motivations discouraged efficiency, placing a greater stress on natural resources. The result was an environmental disaster.

Clean Competition

Severe environmental problems rarely persist in privately owned properties. . . . In formerly socialist Eastern Europe, pollution and environmental destruction occurred at an alarming rate mainly because state-run industries used antiquated, dirty technologies. Those technologies were being phased out long ago in the United States—prior to the creation of the Environmental Protection Agency—because private ownership encouraged the competitive innovation of cleaner technologies.

Washington Times, August 25, 1992.

John Kenneth Galbraith, an avowed proponent of statist economic policies, inadvertently suggested a new approach to environmental protection. In an oft-quoted speech he noted that the U.S. was a nation in which the yards and homes were beautiful and in which the streets and parks were filthy. Galbraith then went on to suggest that we effectively nationalize the yards and homes. For those of us who believe in property rights and economic liberty, the obvious lesson is quite the opposite.

Free market environmentalists seek ways of placing these properties in the care of individuals or groups concerned about their well-being. This approach does not, of course, mean that trees must have legal standing, but rather a call for ensuring that behind every tree, stream, lake, air shed and whale stands one or more owners who are able and willing to protect and nurture that resource.

Consider the plight of the African elephant. On most of the continent, the elephant is managed like the American buffalo once was. It remains a political resource. Elephants are widely viewed as the common heritage of all the peoples of these nations, and are thus protected politically. The "common property" management strategy being used in Kenya and elsewhere in East and Central Africa has been compared and contrasted with the experiences of those nations such as Zimbabwe which have moved decisively in recent years to transfer elephant ownership rights to regional tribal councils. The differences are dramatic. In Kenya, and indeed all of eastern Africa, elephant populations have fallen by over 50 percent in the last decade. In contrast, Zimbabwe's elephant population has been increasing rapidly. As with the beaver in Canada, a program of conservation through use that relies upon uniting the interests of man and the environment succeeds where political management has failed.

The Market and Sustainability

The prophets of sustainability have consistently predicted an end to the world's abundant resources, while the defenders of the free market point to the power of innovation—innovation which is encouraged in the marketplace. Consider the agricultural experience. Since 1950, improved plant and animal breeds, expanded availability and types of agri-chemicals, innovative agricultural techniques, expanded irrigation, and better pharmaceutical products have all combined to spur a massive expansion of world food supplies. That was not expected by those now championing "sustainable development." Lester Brown, in his 1974 Malthusian publication *By Bread Alone*, suggested that crop yield increases would soon cease. Since that date, Asian rice yields have risen nearly 40 percent, an approximate increase of 2.4 percent per year. This rate is similar to that of wheat and other grains. In the developed world it is food surpluses, not food shortages, that present the greater problem, while political institutions continue to obstruct the distribution of food in much of the Third World.

"Profit Incentive" Reduces Waste

Man's greater understanding and ability to work with nature have made it possible to achieve a vast improvement in world food supplies, to improve greatly the nutritional levels of a majority of people throughout the world, in spite of rapid population growth. Moreover, this has been achieved while reducing the stress to the environment. To feed the current world population at current nutritional levels using 1950 yields would require plowing under an additional 10 to 11 million square miles, almost tripling the world's agricultural land demands (now at 5.8 million square miles). This would surely come at the expense of land being used for wildlife habitat and other applications.

Moreover, this improvement in agriculture has been matched by improvements in food distribution and storage, again encouraged by natural market processes and the "profit incentive" that so many environmentalists deplore. Packaging has made it possible to reduce food spoilage, reduce transit damage, extend shelf life, and expand distribution regions. Plastic and other post-use wraps along with the ubiquitous Tupperware have further reduced food waste. As would be expected, the United States uses more packaging than Mexico, but the additional packaging results in tremendous reductions in waste. On average, a Mexican family discards 40 percent more waste each day. Packaging often eliminates more waste than it creates.

Oblivious to Reality

Despite the fact that capitalism has produced more environment-friendly innovations than any other economic system, the advocates of sustainable development insist that this process must be guided by benevolent government officials. That such efforts, such as the United States' synthetic fuels project of the late 1970s, have resulted in miserable failures is rarely considered. It is remarkable how many of the participants at the U.N. Earth Summit seemed completely oblivious to this historical reality.

In the free market, entrepreneurs compete in developing low-cost, efficient means to solve contemporary problems. The promise of a potential profit, and the freedom to seek after it, always provides the incentive to build a better mousetrap, if you will. Under planned economies, this incentive for innovation can never be as strong, and the capacity to reallocate resources toward more efficient means of production is always constrained.

This confusion is also reflected in the latest environmental fad: waste reduction. With typical ideological fervor, a call for increased efficiency in resource use becomes a call to use less of everything, regardless of the cost. Less, we are told, is more in terms of environmental benefit. But neither recycling nor material or energy use reductions *per se* are a good thing, even when judged solely on environmental grounds. Recycling paper often results in increased water pollution, increased energy use, and in the United States, actually discourages the planting of new trees. Mandating increased fuel efficiency for automobiles reduces their size and weight, which in turn reduces their crashworthiness and increases highway fatalities. Environmental policies must be judged on their results, not just their motivations.

Overcoming Scarcity

Environmentalists tend to focus on ends rather than process. This is surprising given their adherence to ecological teaching. Their obsession with the technologies and material usage patterns of today reflects a failure in understanding how the world works. The resources that people need are not chemicals, wood fiber, copper, or the other natural resources of concern to the sustainable development school. We demand housing, transportation, and communication services. How that demand is met is a derivative result based on competitive forces—forces which respond by suggesting new ways of meeting old needs as well as improving the ability to meet such needs in the older ways.

Consider, for example, the fears expressed in the early postwar era that copper would soon be in short supply. Copper was the lifeblood of the world's communication system, essential to link together humanity throughout the world. Extrapolations suggested problems and copper prices escalated accordingly. The result? New sources of copper in Africa, South America, and even the U.S. and Canada were found. That concern, however, also prompted others to review new technologies, an effort that produced today's rapidly expanding fiber optics links.

Such changes would be viewed as miraculous if not now commonplace in the industrialized, and predominantly capitalistic, nations of the world. Data assembled by Lynn Scarlett of the Reason Foundation noted that a system requiring, say, 1,000 tons of copper can be replaced by as little as 25 kilograms of silicon, the basic component of sand. Moreover, the fiber optics system has the ability to carry over 1,000 times the information of the older copper wire. Such rapid increases in communication technology are also providing for the displacement of oil as electronic communication reduces the need to travel and commute. The rising fad of telecommuting was not dreamed up by some utopian environmental planner, but was rather a natural outgrowth of market processes.

It is essential to understand that physical resources are, in and of themselves, largely irrelevant. It is the interaction of man and science that creates resources: Sand and knowledge become fiber optics. Humanity and its institutions determine whether we eat or die. The increase of political control of physical resources and new technologies only increases the likelihood of famine.

"*For environmental politics to confront the real source of environmental degradation . . . it must challenge the prerogatives of private capital.*"

Private Industry Is Bad for the Environment

John Miller

Economist John Miller argues in the following viewpoint that mainstream economics and sound environmental practices are, contrary to statements by self-defined "environmental economists," incompatible. Miller contends that the government should take the initiative in developing environmentally appropriate technologies and should legislate against environmentally damaging practices employed by industry, which he believes is motivated by profit alone. Miller belongs to the collective that produces *Dollars & Sense*, a monthly magazine devoted to socialist interpretation of current economic events, and teaches economics at Wheaton College.

As you read, consider the following questions:

1. How does the market system divert disproportionate levels of environmental degradation onto the world's poorer communities, in Miller's opinion?
2. According to Miller, why are corporations reluctant to develop appropriate technologies?
3. How do orthodox economic calculations place lower value on the future than on the present, according to the author?

John Miller, "The Wrong Shade of Green," *Dollars & Sense*, April 1993. *Dollars & Sense* is a progressive economics magazine published ten times a year. First-year subscriptions cost $16.95 and may be ordered by writing *Dollars & Sense*, One Summer St., Somerville, MA 02143 or calling 617-628-2025.

We are all environmentalists now. A 1992 *Wall Street Journal/*
NBC News Poll revealed that eight in 10 Americans call themselves "environmentalists." The Clinton administration has
anointed Vice President Al Gore, author of the best-seller *Earth
in the Balance,* as its environmental czar. And even George Bush,
our retired "environmental president," has joined up. When he
signed the 1990 Clean Air Act, Bush declared that "every American expects and deserves to breathe clean air."

Surprisingly, Bush's "rights-based" pronouncement lies at the
heart of true environmentalism. An environmentalist philosophy would guarantee pollution standards adequate to protect
the health of all citizens—rich and poor, this generation and future generations—and place economic and ecological concerns
on genuinely equal footing.

A Faustian Bargain

But one group still refuses to wear the environmentalist label:
economists. To orthodox economists, some of whom call themselves environmental economists, environmentalists' claims
about citizens' absolute rights to breathe clean air make little
sense. One can't just say no to pollution, they claim. Pollution, a
"bad," is an inevitable by-product of the same production process that makes economic "goods"; a growing economy means
pollution. And pollution, these economists say, must be managed as the economy is managed—through the market.

Such problematic thinking pervades politics and business.
Politicians and corporate heads not interested in preventing toxic
wastes or developing environmentally friendly products have
seized on this logic to justify their decisions. Judgments made by
the Bush administration's Competitiveness Council are only one
sorry example. The Council, created for corporate interests and
directed by Vice President Dan Quayle, relied on flawed cost-
benefit analyses to argue that the economic costs imposed by the
1990 Clean Air Act exceeded its environmental benefits. The
Council then forced a delay in enacting the legislation.

These bad environmental decisions will not stop with the
Clinton administration. Clinton has abolished the Competitiveness Council but much of its approach to environmental policy
lingers. Surely Clinton's record of sacrificing Arkansas's environment to attract new industry suggests that he too will subordinate environmental policy to economic growth.

The persistent influence of orthodox economics is truly heart-
breaking. It is anchored in the status quo of global inequality
and business as usual. As such, it has become a major obstacle
to developing a sound policy for managing the environment and
the economy.

In 1991, a memo by the World Bank's chief economist Larry

Summers provided an alarming example of how economists subordinate environmental policy to economic growth. In his controversial memo reviewing the bank's *World Development Report*, Summers argued that "the economic logic behind dumping a load of toxic waste in the lowest wage country is impeccable."

To many economists, Summers's logic is indeed impeccable. From their perspective, many developing countries, with limited development and low wages, are "vastly underpolluted." For example, moving a chemical corporation to a developing African nation would bring that country jobs and enlarge its tax base; the accompanying pollution and the risk of cancer and other life-threatening diseases is simply a necessary by-product.

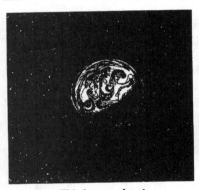
This is your planet

This is your planet on private profits

Coble Howard for *People's Weekly World*. Used with permission.

Economists such as Summers fail to consider the Faustian bargain that developing nations face in the trade-off of pollution for new jobs and taxes. Principally, the pact does not benefit those most in need. The political structure of these countries, much of which remains from colonial eras, means that taxes paid by the polluting corporations will be low, regulation lax, and any benefits will flow to the countries' elites.

The global reality is that these nations confront a world in which they have little access to wealth and resources except through these unequal bargains. The unjust distribution of the world's wealth and resources points to one of several false assumptions central to orthodox economics.

In economists' eyes, the market system gives each economic actor—First World and Third World, producer and consumer, workers and bosses—an equal voice and power. But that, of course, is not true. The richest 20% of the world's people have 150 times the income—and therefore far more power—than the poorest 20%. And the world's wealthiest nations, which have 25% of the world's population, consume 70% of its resources. Not surprisingly, they also spew out most of the world's pollution. The industrial world accounts for up to 90% of the carbon dioxide (CO_2) (mainly from fossil fuel burning) that has so far accumulated in the earth's atmosphere.

Market solutions to environmental problems do not address this inequality; in fact, they play into it. They shove environmental problems under the carpet by shifting pollution and environmental degradation from wealthy to poor areas—both within countries and from the industrial world to the Third World—rather than seeking ways to prevent pollution. The developed world, for example, ships an estimated 20 million tons of waste to the Third World each year. And, in the southern United States, a 1983 study by the U.S. General Accounting Office revealed that three out of four off-site commercial hazardous waste landfills were located in primarily black communities even though blacks represented only 20% of that region's population.

More Ain't Always Better than Less

Economists' logic is flawed in other ways as well. To them, zero pollution is unacceptable because zero growth is unacceptable. What they fail to recognize is that continued economic growth, as we know it, is not an option. There are indeed finite demands humans can put on the earth. Most people who aren't economists understand this. Kenneth Boulding, a gray-beard economist and longtime critic of orthodox economics, once remarked that "anybody who thinks the economy can continue to grow geometrically or indefinitely in a finite world must be a madman, or an economist."

For starters, economists measure the growth of the economy with inaccurate tools. Conventional national income accounts, measured by indicators such as gross domestic product, ignore how growth depletes natural resources and how the loss of those resources compromises the ability of both rich and poor nations to continue to grow. For example, a country could deplete its petroleum reserves, destroy its forests, and erode its soil, thereby irreparably damaging its capacity for future growth, before its national accounts even recognized the problem.

The World Bank and the World Resources Institute, a private environmental think tank, have worked to develop alternative measures that more accurately reflect the relationship between economic growth and the environment. Herman Daly, a World Bank economist, developed the Index of Sustainable Economic

Welfare, an environmentally sensitive measure of economic well-being. It deletes wasteful expenditures, such as the military budget and national advertising; adjusts for the costs of pollution and environmental damage; and accounts for depletion of nonrenewable resources. By Daly's measure, U.S. citizens were no better off in 1986 than they were in 1973—despite real GDP per capita increasing nearly 2% per year.

Green Technology

But even if economic measures accurately capture the trade-off between economic growth and the environment, economists are mistaken to concentrate on such measures exclusively. Barry Commoner, perhaps the best known U.S. environmentalist, argues that the only lasting improvements in environmental quality have come from eliminating pollution-producing technologies. The classic example is lead-free gas, which has dramatically lowered levels of lead pollution. Commoner argues that "since pollution is inherent in the very design of the production technologies, once the technological choice is made, regulation can have only a limited effect, addressing the symptom rather than the disease."

A green economic perspective would focus on developing and using more appropriate technologies, some of which already exist. The *Gaia Atlas of Green Economics* reports that for some industries, zero net pollution technology is a realistic goal. Axel Iveroth, former director-general of the Federation of Swedish Industries, told the Ecology 1989 Congress in Gothenburg that Swedish industry planned to reduce pollution to almost zero by the end of the century.

Adopting appropriate technologies could have dramatic results. In the United States, the Office of Technology Assessment says advanced technologies could slash hazardous industrial waste by 75%. And several U.N. agencies' studies have shown that, with the proper infrastructure, industrial countries could recycle more than 50% of their paper, glass, plastics, and metals.

The environmentalists' concept of appropriate technology has no meaning in standard economic language. According to orthodox economics, the "appropriate" technology is the most profitable. Profitable investments, however, are often not environmentally friendly. Corporations have little incentive to adopt new technologies that reduce environmental costs but that don't improve, or may in fact reduce, profits. Moreover, a corporation is unlikely to invest in developing new technology if other corporations can adopt it without making the investment. So, without government policies that recognize their public benefit, appropriate technologies go undeveloped.

Solar batteries, which use photovoltaic cells, are one classic example of a cost-effective technology the government chose to ignore. Back in 1979, a Federal Energy Administration study predicted that if the U.S. government committed to purchasing 152 million watts of photovoltaic cells—over five times the market then facing U.S. producers—it would allow the industry to develop batteries that it could price competitively in some residential electricity markets by 1984. The government made no such commitment and today experimental solar batteries compete only in high-priced markets.

Horse and Rabbit Stew

Picking appropriate technologies requires taking a long and wide view, as all questions of environmental management do. We must assess both how the current neglect of the environment harms present and future generations, and how adopting new technologies will affect the environment and the economy. But the methods economists use to measure such costs are fraught with problems.

Economists most frequently use a technique called cost-benefit analysis to calculate the price tag of environmental damage. The rationale behind cost-benefit analysis sounds simple and straightforward: No one, or no organization, should implement a policy if anticipated costs exceed anticipated benefits. Economists calculate the costs by assigning monetary values to environmental and human damage. But putting a price tag on global warming, or on human life, is neither straightforward nor simple.

William Nordhaus, a Yale University economist, conducted a cost-benefit study for the Brookings Institution assessing how global warming would harm the U.S. economy. He found that doubling CO_2 concentrations (the chief greenhouse gas) over the next century is "likely to have only a small effect on the U.S. economy." He estimated that by reducing carbon emissions 10%, the United States could produce "optimal" levels of pollution and production. This reduction would limit, but not prevent, global warming.

Not surprisingly, Nordhaus's results horrified many environmental scientists. Paul Ehrlich, a Stanford University biologist, commented that "Nordhaus is so far off base that it's difficult to come to grips with it." As Nordhaus himself points out, his estimates fail to account for several important costs of global warming, including "human health, biological diversity, amenity values of everyday life and leisure, and environmental quality."

That such vital elements of human life are left out has given even mainstream economists pause. One critic, Robert Dorfman, a Harvard economist, likened these cost-benefit analyses to trying to test the flavor of a one-horse and one-rabbit stew by only sampling the rabbit. The horse surely does more to flavor

147

the stew than the rabbit.

When economists do actually try to "measure" those intangible elements of life, they produce equally problematic results. Economist Paul Portney, a senior fellow for Resources for the Future, a Washington, D.C.-based think tank, conducted a cost-benefit study of the 1990 amendments to the Clean Air Act. While many environmentalists hailed the Clean Air Act as the first federal regulation aimed at curbing acid rain, Portney argues that the 1990 legislation's push for greater air quality is not cost effective.

His analysis suggests that the law, once fully phased in, will impose $30 billion in costs on the economy and only generate benefits of $15-20 billion. In Portney's estimation, the regulations forcing companies to reduce air pollutants that harm human health will cost three times their projected benefits. How did Portney calculate this result? He took Environmental Protection Agency estimates that said controlling these hazardous chemicals would prevent 500 cancer deaths per year. He then measured the cost of these lives by using the wage premium coal miners, police, and other workers are paid to face dangerous conditions, an amount other statistical studies place at about $3 million per life.

A life, by most noneconomists' measures, is worth more than the wages necessary to attract workers with few alternatives into dangerous working conditions. Even some Bush administration officials agreed. Robert Brenner, the director of the EPA's Air Policy Office under Bush, noted that "The public has a right to not be exposed to unhealthy levels of air pollution. There are values at stake here other than economics."

The Seventh Generation

One of those values is what type of world we will leave future generations. From a green perspective, the long-term consequences of environmental degradation are often the most horrifying—from global warming to the loss of the rain forest to the extinction of animal species—and the most difficult to assign a monetary value. In addition, the right to a clean environment is no less inviolable for future generations than for today's citizens. Environmentalists often invoke the directive of the Iroquois nation that requires its tribal councils to consider how their decisions will affect the seventh generation into the future, approximately 150 years later.

Environmental rights of future generations don't cut much ice with mainstream economists, however. Larry Summers, blunt as always, told the *Economist* that "the argument that a moral obligation to future generations demands special treatment of environmental investments is fatuous." As he sees it, "we can help our descendants as much by improving infrastructure as by preserving rain forests."

The problems with Summers's conclusion begin with his comparison. For Summers, rebuilding highways is but one alternative use of the funds necessary to save the rain forest. To justify protecting the rain forest, there must be conventionally measured economic gains that exceed those of all alternative investments. This logic ignores the fact that you can replace, or rebuild, a highway; you cannot replace or rebuild the rain forest. The rain forest—with all its biological richness—has inestimable value.

Economists, however, believe they can measure such elusive costs. When calculating the future benefits of addressing an environmental problem today, economists discount the future's value to reflect both the importance society places on the present and the possible alternative uses of these funds.

The choice of the discount rate—which reflects the rates of return of other investments—is crucial in these calculations. For example, an economist may be estimating the effects of a project to address global warming over the next 200 years. A high discount rate might be 7%. That rate would convert $1 million of environmental damage in 200 years to a present value of $1.33. Even a lower discount rate of 1.5%—based on projected per capita growth rates rather than on a return on an investment—would only ring up $51,000 for global warming damage. As E.J. Mishan, the famed British cost-benefit economist, has argued, environmental costs that affect future generations should not be discounted at low or high rates. Future generations are both absent from these environmental decisions and place more importance on the future (their present) than today (their past).

The Right Shade of Green

A progressive environmental policy starts with a non-negotiable demand for population standards adequate to protect both present and future human health and quality of life. This demand is no different than the demand for universal access to basic health care or decent housing. In this context, the "optimal" level of pollution is no more relevant to the policy debate than the "optimal" level of mental health.

By focusing on why we want a growing and productive economy—to protect our welfare and the welfare of our children—environmentalism teaches an important lesson: Protecting the environment is a much better investment than conventional economics would lead us to believe.

How can we better recognize the value of environmental investments rejected by traditional economic reasoning? Some environmentalists have developed environmentally sensitive investment rules. The World Bank's Daly favors a "no net loss" of natural capital rule. That would preclude projects that destroy forests, drain wetlands, dam rivers, or pave over crop lands unless the project also replenishes a compensating resource.

More generally, an industrial policy that directs investment by a democratic calculus, not an economic one, and places as much importance on long-term ecology as on short-term profitability remains the heart and soul of a green strategy. Some market-based policies could help here. For instance, Robert Repetto of the World Resources Institute would legislate a series of "green fees"—for example, a carbon tax to reduce CO_2 emissions—intended to alter the economic trade-off between output and pollution. Significantly higher taxes on pollution, Repetto argues, would curb environmentally destructive corporate behavior.

But market-based fees and taxes alone will not stimulate the widespread adoption of alternative technologies that we need. Fundamental changes in how and what we produce will require banning some dangerous chemicals, and this will inevitably mean some industrial workers will lose their jobs. The environmental movement must recognize the public's obligation to aid these workers.

Facing up to the demands an ecologically saner world will make on the economy will not be easy. It will mean taking on greater costs and confronting the politicians and corporations who use orthodox economics to justify continued neglect of the environment.

For environmental politics to confront the real source of environmental degradation—the use of destructive technologies—it must challenge the prerogatives of private capital. To do this, we will have to reject the common sense of traditional economics in favor of a common sense that is red as well as green.

Periodical Bibliography

The following articles have been selected to supplement the diverse views presented in this chapter. Addresses are provided for periodicals not indexed in the *Readers' Guide to Periodical Literature*, the *Alternative Press Index*, or the *Social Sciences Index*.

Business and Society Review	"America's Worst Toxic Polluters," Winter 1993. Available from 25-13 Old Kings Highway N., Suite 107, Darien, CT 06820.
Ruth N. Caplan	"Growing the Economy Green," *Christian Social Action*, March 1993. Available from 100 Maryland Ave. NE, Washington, DC 20002.
Don J. DeBenedictis	"Few Like Pollution Guidelines," *ABA Journal*, June 1993. Available from 750 N. Lake Shore Dr., Chicago IL 60611.
John R. Ehrenfeld	"Designing Green Goods," *World & I*, April 1995. Available from 3600 New York Ave. NE, Washington, DC 20002.
John Bellamy Foster	"'Let Them Eat Pollution': Capitalism and the World Environment," *Monthly Review*, January 1993.
Paul Hawken, interviewed by Allan Hunt Badiner	"Natural Capitalism," *Yoga Journal*, September/October 1994. Available from PO Box 469018, Escondido, CA 92046-9018.
Beatriz Johnston Hernandez	"Dirty Growth," *New Internationalist*, August 1993.
Art Kleiner	"The Ant, the Grasshopper, and the GNP," *Garbage*, February/March 1993.
Christina Nichols	"The Greenwashing of America," *Other Side*, November/December 1994. Available from 300 W. Apsley St., Philadelphia, PA 19144-4285.
Robert Repetto	"Trade Can Serve the Environment," *Amicus Journal*, Fall 1993.
Richard A. Rosen	"Reclaiming Economics," *Dollars & Sense*, July/August 1993.
Julian Szekely and Gerardo Trapaga	"From Villain to Hero," *Technology Review*, January 1995.

How Should the Environment Be Protected?

The Environment

Chapter Preface

Few, if any, will deny that the technological explosion of the twentieth century revolutionized the way human society functions. In doing so, it dramatically altered how human beings interact with the earth. "Technophiles," or those who sing the praises of technology, count among its many blessings the longer and healthier lives and improved standards of living humans as a whole are enjoying.

These individuals foresee technology as tomorrow's answer to whatever problems remain unsolved today, including those that affect the health of the planet. For example, according to Hunter S. Lovins, cofounder of the Rocky Mountain Institute, current U.S. oil consumption can be reduced by four-fifths with technology that already exists. Indur M. Goklany, who manages the science and engineering branch of the U.S. Department of the Interior, also sees technological advances as humankinds' greatest hope for a sustainable future. "Increases in the productivity of agriculture and forestry over the last few decades have done more to slow land conversion, and preserve habitat . . . than all the other protection measures put together," he contends.

Not everyone views technology as the key to sustaining a healthy environment. For every "technophile" there appears to be a "technophobe"—someone who fears and distrusts technology. These individuals note that not all technologies are good for the environment, emphasizing that some are actually quite destructive. "What is important," writes Robert Gilman, editor of *In Context* magazine, "is . . . the *specific* environmental impacts that particular technologies have." He reminds readers of technologies like "slash-and-burn" agriculture, which contributes to the depletion of tropical rain forests, and nuclear power, which produces toxic wastes. Newspaper columnist Molly Ivins also advocates a critical stance in relation to technology. "In the name of progress and technology, we have harmed the ecology to an extent that kills the goose that lays the golden eggs," she argues.

While some are concerned with the dark side of technology and others with its potential benefits, many people agree that technology itself is neither inherently good nor bad. They argue that the responsibility for whatever results it leads to ultimately lies in the hands of those who use it. New technologies represent only one of many measures proposed for protecting the environment. The authors included in this chapter debate a variety of approaches to preserve the earth.

> *"In order for a sustainable society to exist, every purchase must reflect . . . the costs to the air, water, and soil; the cost to future generations."*

Redesigning Industry for Sustainability Will Protect the Environment

Paul Hawken

"Sustainable development" has become an increasingly popular concept among environmentalists in recent years. To its proponents, "sustainability" means human consumption of the earth's resources at the same rate nature produces or replenishes them. In the following viewpoint, Paul Hawken argues that current systems of production and commerce must be redesigned to achieve sustainable rates of consumption if the global human community and the natural world are to survive. He suggests enforcing greater controls over corporations to prevent industrial pollution, adjusting the price of goods to include the environmental and human costs of production, and adapting methods of production to the cycles of nature. This viewpoint presents ideas adapted from Hawken's book *The Ecology of Commerce*.

As you read, consider the following questions:

1. How was the relationship between corporations and individual citizens originally envisioned, according to Hawken?
2. How does Hawken explain the categories "consumables," "durables," and "unsalables?"

Excerpted from "A Declaration of Sustainability" by Paul Hawken, *Utne Reader*, September/ October 1993. Reprinted by permission.

In order to approximate a sustainable society, we need to describe a system of commerce and production in which each and every act is inherently sustainable and restorative. Because of the way our system of commerce is designed, businesses will not be able to fulfill their social contract with the environment or society until the system in which they operate undergoes a fundamental change, a change that brings commerce and governance into alignment with the natural world from which we receive our life. . . .

What is needed is a conscious plan to create a sustainable future, including a set of design strategies for people to follow. For the record, I will suggest several.

Take Back the Charter

Although corporate charters may seem to have little to do with sustainability, they are critical to any long-term movement toward restoration of the planet. Read *Taking Care of Business: Citizenship and the Charter of Incorporation*, a 1992 pamphlet by Richard Grossman and Frank T. Adams. In it you find a lost history of corporate power and citizen involvement that addresses a basic and crucial point: Corporations are chartered by, and exist at the behest of, citizens. Incorporation is not a right but a privilege granted by the state that includes certain considerations such as limited liability. Corporations are supposed to be under our ultimate authority, not the other way around. The charter of incorporation is a revocable dispensation that was supposed to ensure accountability of the corporation to society as a whole. When Rockwell criminally despoils a weapons facility at Rocky Flats, Colorado, with plutonium waste, or when any corporation continually harms, abuses, or violates the public trust, citizens should have the right to revoke its charter, causing the company to disband, sell off its enterprises to other companies, and effectively go out of business. The workers would have jobs with the new owners, but the executives, directors, and management would be out of jobs, with a permanent notice on their résumés that they mismanaged a corporation into a charter revocation. This is not merely a deterrent to corporate abuse but a critical element of an ecological society because it creates feedback loops that prompt accountability, citizen involvement, and learning. We should remember that the citizens of this country originally envisioned corporations to be part of a public-private partnership, which is why the relationship between the chartering authority of state legislatures and the corporation was kept alive and active. They had it right.

The economy is environmentally and commercially dysfunctional because the market does not provide consumers with proper information. The "free market" economies that we love so much are excellent at setting prices but lousy when it comes to recognizing costs. In order for a sustainable society to exist, every purchase must reflect or at least approximate its actual cost, not only the direct cost of production but also the costs to the air, water, and soil; the cost to future generations; the cost to worker health; the cost of waste, pollution, and toxicity. Simply stated, the marketplace gives us the wrong information. It tells us that flying across the country on a discount airline ticket is cheap when it is not. It tells us that our food is inexpensive when its method of production destroys aquifers and soil, the viability of ecosystems, and workers' lives. Whenever an organism gets wrong information, it is a form of toxicity. In fact, that is how pesticides work. A herbicide kills because it is a hormone that tells the plant to grow faster than its capacity to absorb nutrients allows. It literally grows itself to death. Sound familiar? Our daily doses of toxicity are the prices in the marketplace. They are telling us to do the wrong thing for our own survival. They are lulling us into cutting down old-growth forests on the Olympic Peninsula for apple crates, into patterns of production and consumption that are not just unsustainable but profoundly shortsighted and destructive. It is surprising that "conservative" economists do not support or understand this idea, because it is they who insist that we pay as we go, have no debts, and take care of business. Let's do it. . . .

Allow Resource Companies to Be Utilities

An energy utility is an interesting hybrid of public-private interests. A utility gains a market monopoly in exchange for public control of rates, open books, and a guaranteed rate of return. Because of this relationship and the pioneering work of Amory Lovins, we now have markets for "negawatts." It is the first time in the history of industrialism that a corporation has figured out how to make money by selling the absence of something. Negawatts are the opposite of energy: They represent the collaborative ability of a utility to harness efficiency instead of hydrocarbons. This conservation-based alternative saves ratepayers, shareholders, and the company money—savings that are passed along to everyone. All resource systems, including oil, gas, forests, and water, should be run by some form of utility. There should be markets in negabarrels, negatrees, and negacoal. Oil companies, for example, have no alternative at present other than to lobby for the absurd, like drilling in the Arctic National Wildlife Refuge. That project, a $40 billion to $60 billion investment for a hoped-for supply of oil that would meet U.S. consumption needs for only six months, is the only way an oil company can make money under our current system of commerce. But what if the oil companies formed an oil utility and cut a deal with citizens and taxpayers that allowed them to "invest" in insulation, super-glazed windows, conservation rebates on new automobiles, and the scrapping of old cars? Through Green fees, we would pay them back a return on their conservation investment equal to what utilities receive, a rate of return that would be in accord with how many barrels of oil they save, rather than how many barrels they produce. Why should they care? Why should we? A $60 billion investment in conservation will yield, conservatively, four to ten times as much energy as drilling for oil. Given Lovins' principle of efficiency extraction, try to imagine a forest utility, a salmon utility, a copper utility, a Mississippi River utility, a grasslands utility. Imagine a system where the resource utility benefits from conservation, makes money from efficiency, thrives through restoration, and profits from sustainability. It is possible today.

Change Linear Systems to Cyclical Ones

Our economy has many design flaws, but the most glaring one is that nature is cyclical and industrialism is linear. In nature, no linear systems exist, or they don't exist for long because they exhaust themselves into extinction. Linear industrial systems take resources, transform them into products or services, discard waste, and sell to consumers, who discard more waste when they have consumed the product. But of course we don't consume TVs, cars, or most of the other stuff we buy. Instead, Americans produce six times their body weight every week in hazardous and toxic waste water, incinerator fly ash, agricultural wastes, heavy metals, and waste chemicals, paper, wood, etc. This does not include CO_2 which if it were included would double the amount of waste. Cyclical means of production are designed to imitate natural systems in which waste equals food for other forms of life, nothing is thrown away, and symbiosis replaces competition. Bill McDonough, a New York architect who has pioneered environmental design principles, has designed a system to retrofit every window in a major American city. Although it still awaits final approval, the project is planned to go like this: The city and a major window manufacturer form a joint venture to produce energy-saving super-glazed windows in the town. This partnership company will come to your house or business, measure all windows and glass doors, and then replace them with windows with an R-8 to R-12 energy-efficiency rating within 72 hours. The windows will have the same casements, molding, and general appearance as the old ones. You will receive a $500 check upon installation, and you will pay for the new windows over a 10- to 15-year period in your utility or tax bill. The total bill is less than the cost of the energy the windows will save. In other words, the windows will cost the home or business owner nothing. The city will pay for them initially with industrial development bonds. The factory will train and employ 300 disadvantaged people. The old windows will be completely recycled and reused, the glass melted into glass, the wooden frames ground up and mixed with recycled resins that are extruded to make the casements. When the city is reglazed, the residents and businesses will pocket an extra $20 million to $30 million every year in money saved on utility bills. After the windows are paid for, the figure will go even higher. The factory, designed to be transportable, will move to another city: the first city will retain an equity interest in the venture. McDonough has designed a win-win-win-win-win system that optimizes a number of agendas. The ratepayers, the homeowners, the renters, the city, the environment, and the employed all thrive because they are "making" money from efficiency rather than exploitation. It's a little like running the industrial economy backwards.

Transform the Making of Things

We have to institute the Intelligent Product System created by Michael Braungart of the EPEA (Environmental Protection Encouragement Agency) in Hamburg, Germany. The system recognizes three types of products. The first are *consumables*, products that are either eaten, or, when they're placed on the ground, turn into dirt without any bio-accumulative effects. In other words, they are products whose waste equals food for other living systems. At present, many of the products that should be "consumable," like clothing and shoes, are not. Cotton cloth contains hundreds of different chemicals, plasticizers, defoliants, pesticides, and dyes; shoes are tanned with chromium and their soles contain lead; neckties and silk blouses contain zinc, tin, and toxic dye. Much of what we recycle today turns into toxic by-products, consuming more energy in the recycling process than is saved by recycling. We should be designing more things so that they can be thrown away—into the compost heap. Toothpaste tubes and other non-degradable packaging can be made out of natural polymers so that they break down and become fertilizer for plants. A package that turns into dirt is infinitely more useful, biologically speaking, than a package that turns into a plastic park bench. Heretical as it sounds, designing for decomposition, not recycling, is the way of the world around us.

The second category is *durables*, but in this case, they would not be sold, only licensed. Cars, TVs, VCRs, and refrigerators would always belong to the original manufacturer, so they would be made, used, and returned within a closed-loop system. This is already being instituted in Germany and to a lesser extent in Japan, where companies are beginning to design for disassembly. If a company knows that its products will come back someday, and that it cannot throw anything away when they do, it creates a very different approach to design and materials.

No Future

Sustainable development really comes down to common-sense economics. . . .

On a global scale, economists and engineers may be confident that free markets will always find substitutes as resources are depleted. But many rural towns and even some big cities may have no future unless governments start taking better care of the surrounding environment and the sustenance it provides them. Not every place needs a thousand-year plan, but many could use public leaders willing to look beyond the next election campaign and start shaping more lasting communities.

Tom Arrandale, *Governing*, April 1994.

Last, there are *unsalables*—toxins, radiation, heavy metals, and chemicals. There is no living system for which these are food and thus they can never be thrown away. In Braungart's Intelligent Product System, unsalables must always belong to the original maker, safeguarded by public utilities called "parking lots" that store the toxins in glass-lined barrels indefinitely, charging the original manufacturers rent for the service. The rent ceases when an independent scientific panel can confirm that there is a safe method to detoxify the substances in question. All toxic chemicals would have molecular markers identifying them as belonging to their originator, so that if they are found in wells, rivers, soil, or fish, it is the responsibility of the company to retrieve them and clean up. This places the problem of toxicity with the makers, where it belongs, making them responsible for full-life-cycle effects.

Vote, Don't Buy

Democracy has been effectively eliminated in America by the influence of money, lawyers, and a political system that is the outgrowth of the first two. While we can dream of restoring our democratic system, the fact remains that we live in a plutocracy—government by the wealthy. One way out is to vote with your dollars, to withhold purchases from companies that act or respond inappropriately. Don't just avoid buying a Mitsubishi automobile because of the company's participation in the destruction of primary forests in Malaysia, Indonesia, Ecuador, Brazil, Bolivia, Canada, Chile, Siberia, and Papua New Guinea.

Write and tell them why you won't. Engage in dialogue, send one postcard a week, talk, organize, meet, publish newsletters, boycott, patronize, and communicate with companies like General Electric. Educate non-profits, organizations, municipalities, and pension funds to act affirmatively, to support the ecological CERES (formerly *Valdez*) Principles for business, to invest intelligently, and to *think* with their money, not merely spend it. Demand the best from the companies you work for and buy from. You deserve it and your actions will help them change.

Restore the "Guardian"

There can be no healthy business sector unless there is a healthy governing sector. In her book *Systems of Survival*, author Jane Jacobs describes two overarching moral syndromes that permeate our society: the commercial syndrome, which arose from trading cultures, and the governing, or guardian, syndrome that arose from territorial cultures. The guardian system is hierarchical, adheres to tradition, values loyalty, and shuns trading and inventiveness. The commercial system, on the other hand, is based on trading, so it values trust of outsiders, innovation, and future thinking. Each has qualities the other lacks. Whenever the guardian tries to be in business, as in Eastern Europe, business doesn't work. What is also true, but not so obvious to us, is that when business plays government, governance fails as well. Our guardian system has almost completely broken down because of the money, power, influence, and control exercised by business and, to a lesser degree, other institutions. Business and unions have to get out of government. We need more than campaign reform: We need a vision that allows us all to see that when Speaker of the House Tom Foley exempts the aluminum industry in his district from the proposed Btu tax, or when Philip Morris donates $200,000 to the Jesse Helms Citizenship Center, citizenship is mocked and democracy is left gagging and twitching on the Capitol steps. The irony is that business thinks that its involvement in governance is good corporate citizenship or at least is advancing its own interests. The reality is that business is preventing the economy from evolving. Business loses, workers lose, the environment loses.

Shift from Electronic Literacy to Biologic Literacy

That an average adult can recognize one thousand brand names and logos but fewer than ten local plants is not a good sign. We are moving not to an information age but to a biologic age, and unfortunately our technological education is equipping us for corporate markets, not the future. Sitting at home with virtual reality gloves, 3D video games, and interactive cable TV shopping is a barren and impoverished vision of the future. The computer revolution is not the totem of our future, only a tool. Don't get me wrong. Computers are great. But they are not an uplifting or compelling vision for culture or society. They do not move us toward a sustainable future any more than our obsession with cars and televisions provided us with newer definitions or richer meaning. We are moving into the age of living machines, not, as Corbusier noted, "machines for living in." The Thomas Edison of the future is not Bill Gates of Microsoft, but John and Nancy Todd, founders of the New Alchemy Institute, a Massachusetts design lab and think tank for sustainability. If the Todds' work seems less commercial, less successful, and less glamorous, it is because they are working on the real problem—how to live—and it is infinitely more complex than a microprocessor. Understanding biological processes is how we are going to create a new symbiosis with living systems (or perish). What we can learn on-line is how to model complex systems. It is computers that have allowed us to realize how the synapses in the common sea slug are more powerful than all of our parallel processors put together.

Take Inventory

We do not know how many species live on the planet within a factor of ten. We do not know how many are being extirpated. We do not know what is contained in the biological library inherited from the Cenozoic age. (Sociobiologist E.O. Wilson estimates that it would take 25,000 person-years to catalog most of the species, putting aside the fact that there are only 1,500 people with the taxonomic ability to undertake the task.) We do not know how complex systems interact—how the transpiration of the giant lily, *Victoria amazonica*, of Brazil's rainforests affects European rainfall and agriculture, for example. We do not know what happens to 20 percent of the CO_2 that is off-gassed every year (it disappears without a trace). We do not know how to calculate sustainable yields in fisheries and forest systems. We do not know why certain species, such as frogs, are dying out even in pristine habitats. We do not know the long-term effects of chlorinated hydrocarbons on human health, behavior, sexuality, and fertility. We do not know what a sustainable life is for existing inhabitants of the planet, and certainly not for future populations. (A Dutch study calculated that your fair share of air travel is one trip across the Atlantic in a lifetime.) We do not know how many people we can feed on a sustainable basis, or what our diet would look like. In short, we need to find out what's here, who has it, and what we can or can't do with it.

The environmental and socially responsible [business] movements would gain additional credibility if they recognized that the greatest amount of human suffering and mortality is caused by environmental problems that are not being addressed by environmental organizations or companies. Contaminated water is killing a hundred times more people than all other forms of pollution combined. Millions of children are dying from preventable diseases and malnutrition.

The movement toward sustainability must address the clear and present dangers that people face worldwide, dangers that ironically increase population levels because of their perceived threat. People produce more children when they're afraid they'll lose them. Not until the majority of the people in the world, all of whom suffer in myriad preventable yet intolerable ways, understand that environmentalism means improving their lives directly will the ecology movement walk its talk. . . .

Respect the Human Spirit

If hope is to pass the sobriety test, then it has to walk a pretty straight line to reality. Nothing written, suggested, or proposed here is possible unless business is willing to integrate itself into the natural world. It is time for business to take the initiative in a genuinely open process of dialogue, collaboration, reflection, and redesign. "It is not enough," writes Jeremy Seabrook of the British Green party, "to declare, as many do, that we are living in an unsustainable way, using up resources, squandering the substance of the next generation however true this may be. People must feel subjectively the injustice and unsustainability before they will make a more sober assessment as to whether it is worth maintaining what is, or whether there might not be more equitable and satisfying ways that will not be won at the expense either of the necessities of the poor or of the wasting fabric of the planet."

Poet and naturalist W.S. Merwin (citing Robert Graves) reminds us that we have one story, and one story only, to tell in our lives. We are made to believe by our parents and businesses, by our culture and televisions, by our politicians and movie stars that it is the story of money, of finance, of wealth, of the stock portfolio, the partnership, the country house. These are small, impoverished tales and whispers that have made us restless and craven; they are not stories at all. As author and garlic grower Stanley Crawford puts it, "The financial statement must finally give way to the narrative, with all its exceptions, special cases, imponderables. It must finally give way to the story, which is perhaps the way we arm ourselves against the next and always unpredictable turn of the cycle in the quixotic dare that is life; across the rock and cold of lifelines, it is our seed, our clove, our filament cast toward the future." It is something deeper than anything commercial culture can plumb, and it is waiting for each of us.

> *"Behind its pro-growth, pro-man veneer, sustainable development holds the same basic ideas as standard environmentalism: . . . production as destruction."*

Sustainable Development Is Impracticable

Bennett C. Karp

Bennett C. Karp is a systems engineer in the telecommunications industry who strongly opposes the idea of sustainable development. Insisting that sustainable development is not possible, Karp argues in the following viewpoint that supporters of this policy are dishonestly representing themselves as pro-growth when in fact they desire no growth whatsoever. He contends that the measures proposed to achieve sustainability endanger individual liberties and threaten to transform capitalist society into one of government-imposed socialism.

As you read, consider the following questions:

1. In what ways is solar electricity "unsustainable," as explained in this viewpoint?
2. Why, in Karp's opinion, is the scarcity of resources "not a problem"?
3. Describe the similarities and differences between sustainable development and standard environmentalism as presented by Karp.

Bennett C. Karp, "Sustaining the Assault on Development (Part II)," *Intellectual Activist*, November 1993. Reprinted with permission.

The advocates of sustainable development attempt to distinguish activities that deplete nature's reserves from activities that are within the "earth's ability to renew itself." This alleged distinction between "sustainable" and "unsustainable" production, between "renewable" and "nonrenewable" resources, is premised on a false view of production—and, beneath that, on a false theory of value.

No Such Thing

In reality, there is no such thing as what they call "sustainable" production. In *all* production, man reshapes nature. In every act of production, he "uses up" some part of nature to create the values that his life requires. Even, for example, in the generation of solar electricity, a supposedly "sustainable" technology, man must manufacture solar cells, build frames for the solar arrays from steel and glass, use vast areas of land for the arrays, establish a transmission system to distribute the electricity. All of the manufactured goods require factories, energy, and raw materials such as silicon, rare metals, iron, and coal. In every aspect of the project, man takes from nature. In this respect, there is no essential difference between solar energy and electricity generated by burning fossil fuels. Thus, to oppose "unsustainable" activities is to oppose all production.

Sustainable development—and all of environmentalism—has a Garden of Eden notion of production. Its view is that man simply gobbles up nature's bounty. It blanks out the fact that it takes work to transform iron ore in the ground into steel, and steel into automobiles. It blanks out the fact that when man has expended that effort, he has created something of value that wasn't there before. To sustainable development, the only thing that has occurred is that man has depleted nature's resources; to it, production is a process of consumption. And since the resources once taken from nature are no longer part of nature, this alleged consumption is seen as impoverishment.

Devaluing the Human Mind

To sustainable development, the human mind plays no essential role in production. But, according to Ayn Rand, "production is the application of reason to the problem of survival." The fundamental "resource" on which production depends is man's mind. *That* resource is always "renewable." Because there are no limits to man's knowledge and ingenuity, there are no limits to production.

As the fundamental resource, man's mind is the answer to any "shortage" of natural resources. As George Reisman has noted, "The entire earth, from the uppermost limits of its atmosphere to its very center, four thousand miles down, consists exclusively of natural resources, of *solidly packed natural resources.*" It is up to man to discover how to exploit them. As technology improves, the resources available to and usable by man *increase.* The scarcity of a particular resource is possible, but not a problem—not, that is, under capitalism. A free market balances supply and demand with the price mechanism. If a resource becomes scarce, its price rises, which simultaneously reduces demand, makes alternative sources of the resource economically feasible, and speeds the search for alternatives. In this way, nineteenth-century "shortages" of wood and whale oil became non-problems.

Since each human mind can improve productivity and since each person is a potential producer, the arguments against population growth are groundless. As technology improves, larger populations can be supported. Larger populations, under capitalism, are both a result of progress and a cause of further progress; they lead to a greater division of labor and higher standards of living for all.

Better Methods, Not Balance

Sustainable development argues that "unsustainable" economic activity will make future production increasingly difficult and ultimately lead to collapse. In fact, every act of production leads to an improvement in *man's* environment and facilitates future production. The key to improved future production is ever-increasing human knowledge. The existential effect of this knowledge is technological progress and capital accumulation, such as the discovery of better methods of production or the building of new factories. These are the economic effects of man's applying a rational, long-term approach to the problem of production. It is man's desire to improve his life that leads him to seek better methods of production. The arguments made by sustainable development amount to the cry that man should not seek to better his life, that he must restrain production so as to remain in "balance" with nature.

The solution to any actual undesirable effects of production and technology is not to stop production and technology, not to seek a "balance," but to seek better methods through improved technology. Advancements in technology have always improved man's environment—by eliminating diseases, replacing coal-generated electricity with nuclear power, increasing food production, etc.

Sustainable development holds its false view of production because, like standard environmentalism, it treats nature as an intrinsic value, i.e., as possessing value apart from its usefulness to man. It is this view of value that sustainable development tries to hide behind its pro-man veneer. Sustainable development regards production as the destruction of intrinsic value, not as the creation of objective value. Its notion of so-called "natural capital" illustrates its view of nature as an intrinsic value. If, for instance, man takes oil from the ground and refines it into heating oil or gasoline, he has in fact produced a value—an *objective* value. To sustainable development, however, oil in the ground is an *intrinsic* value, the only kind of "value" it recognizes. So, it claims, the oil production represents the destruction of value. Only activities that are "sustainable" are considered non-destructive and thus acceptable. But neither oil in the ground nor any other resources in their natural state, if declared off limits to man, are values. Properly, such resources are *potential* values, which man can bring into existence through production.

Sustainable Delusions

Sustainable development . . . can best be understood as a euphemism for environmental socialism—granting governments more and more control over the allocation of resources in the name of environmental protection. But if any lesson can be learned from the collapse of socialism in the former communist countries, it is that government ownership and control of resources is a recipe for economic collapse and environmental degradation. . . .

Private property, free markets, and sound liability laws—anathemas to a theory of sustainable development—are essential for a clean environment and for economic growth.

Thomas J. DiLorenzo, *The Futurist*, September/October 1993.

Natural resources are a value, sustainable development decrees, not because of their value to man, but because they are a value in and of themselves. If, however, one grasps that value is objective, i.e., that value is a relationship between an object or goal in reality and the valuer, that value as Ayn Rand puts it "presupposes an answer to the question: of value to *whom* and for *what*?" and that for man, value is measured by the standard of man's life—then the whole issue of nature as a value in itself apart from man evaporates.

Demanded Sacrifice for the Unborn

The assertion that we owe it to future generations to save something for them—such as the call from a group established by the U.N. [United Nations] for "intergenerational equity," whereby decisions by the present generation are made with an awareness of their impact on future generations—is another form in which sustainable development expresses the environmentalist doctrine of intrinsic value. Consistent with sustainable development's basic approach, it is a form designed to provide a pro-man veneer to environmentalism's anti-man ideology. Instead of explicitly claiming that nature has value in itself, we are told that it has value to "future generations." But in both cases, the argument is the same: there is something beyond man that has value apart from its relation to his life—and the practical demand is the same: sacrifice. Going deeper than more common versions of altruism, this metaphysical perversion gives "equity" to the not-yet-born. It is a re-packaging of socialism's call for sacrifice for the future. Just as socialism tells its subjects that they must toil without reward so that there may be prosperity in the "future," sustainable development demands that we must "limit affluence" for the alleged benefit of "future generations." And just as prosperity never arrives under socialism, future generations will be told under sustainable development that *they* have to sacrifice for their future generations.

In fact, there is no conflict between our own interests and the well-being of the men of future generations. Consider where *we* would be if previous generations, instead of developing the modern technology that has given us our advanced standard of living, had tried to "balance" progress with "ecological concerns." Who made the discoveries and created the wealth that has enabled us to enjoy today's high technology, longer life spans, and other modern wonders: is it the giants of the Industrial Revolution or the men of the Dark and Middle Ages who lived "sustainably" for a millennium?

Beneath the Veneer

Sustainable development's demand that economics and accounting include "environmental costs" illustrates its false views of production and of value. The "environmental costs" championed by sustainable development are alleged costs that are supposedly above and beyond the actual costs of production. The only relevant costs, i.e., the only ones that need to be accounted for, are the costs to man. For example, a company that mines iron counts as its costs the depreciation on the mine and all the costs of extracting the ore. At every stage of production, all the costs *to man* are counted. But sustainable development demands that some additional cost, the alleged cost *to nature*, be counted, e.g., that beyond the depreciation on the mine, there is some other cost since the ore is no longer in the ground. These "costs" represent nothing more than the intrinsic value that environmentalists claim for nature. Since there are no such costs to measure, under sustainable development, economics and accounting *do* become tools of distortion.

Behind its pro-growth, pro-man veneer, sustainable development holds the same basic ideas as standard environmentalism: nature as an intrinsic value and production as destruction. . . .

Sustainable development, in spite of its pro-growth veneer, shares standard environmentalism's philosophic opposition to the material world. Noting that "throughout the ages, philosophers and religious leaders have denounced materialism as a path to human fulfillment," the Worldwatch Institute states that "materialism simply cannot survive the transition to a sustainable world." We will have "to adopt simpler and less consumptive lifestyles" to the extent that "it will become unfashionable to own fancy new cars and clothes and the latest electronic devices."

A Socialist Agenda

Environmentalism and socialism, with their shared antipathy to capitalism, have always been allies. Under its guise of protecting future economic activity, sustainable development further integrates the socialist agenda into environmentalism. Sustainable development demands massive aid from the industrial nations to the undeveloped countries for environmental protection. Borrowing socialism's line that "a hungry man can't be free," sustainable development argues, in the words of R.K. Pauchari, "A person who is worrying about his next meal is not going to listen to lectures on protecting the environment." It is supposedly in *our interest* to give away our wealth to the Third World. "We can no longer separate the future habitability of the planet from the current distribution of wealth," says Lester Brown, president of the Worldwatch Institute.

The industrialized nations must, in effect, pay extortion money to the undeveloped countries to keep them from harming the environment. As a Western diplomat explains: "The Chinese [as well as other poor countries] have made the choice that if there is a decision to be made between development and environment, they'll go for development. If the West wants a better environment, they'll have to pay." According to sustainable development, if the poor countries destroy their environment, *we* suffer. Thus, we must pay them *not* to develop. To draw an analogy to another foreign-aid debacle, while it is a complete waste or worse to bribe the Russians to adopt capitalism, this environmental blackmail is like paying the Russians to return to communism.

Underlying this similarity between sustainable development and socialism is egalitarianism. Socialism's version of egalitarianism is that the productive should be sacrificed to the nonproductive. Standard environmentalism's version is that man should be sacrificed to nature. Sustainable development's integration of the two is that the productive should be sacrificed to the non-productive in the name of nature. Behind the rational-ization of "women's rights" as a means to stabilizing population, it is sustainable development's egalitarian viewpoint that led to the inclusion of women's rights (and aid to undeveloped coun-tries) in the agenda of the Earth Summit.

In short, all these socialist programs are consistent with sus-tainable development because they represent restrictions on production. According to sustainable development, all restric-tions on production help "sustain" the environment. The ideal of sustainable development is *non*-production.

With its veneer of being long-range and pro-man, with its no-tion of "balance," sustainable development makes environmen-talism appear *practical*. It brands its opponents as unconcerned with life beyond the immediate moment thus reinforcing envi-ronmentalism's *moral sanction*. Like standard environmentalism, sustainable development seeks drastically to curtail man's activ-ities. Both demand sacrifice; both call for more governmental vi-olation of individual rights. Both deny the reality of production; both uphold the intrinsic value of nature. The difference is this: with its pro-man, pro-production veneer, sustainable develop-ment has a greater ability to garner support than does standard environmentalism. . . .

Man's Life as a Standard

Sustainable development represents a pragmatic attempt to combine opposite principles. There is no possibility of "balance" or compromise between opposite principles. The good element has nothing to gain from the bad element. The calls for "bal-ance" are attempts to propagate evil by sucking the blood of the life-sustaining good. There is no such thing as "too much good."

As Leonard Peikoff has noted:

> Just as there cannot be too much rationality, so there cannot be too much of any of its derivatives, including productive-ness. Just as there is no limit to man's need of knowledge and therefore of thought, so there is no limit to man's need of wealth and therefore of creative work. Intellectually, every dis-covery contributes to human life by enhancing men's grasp of reality. Existentially, every material achievement contributes to human life by making it increasingly secure, prolonged, and/or pleasurable. There can be no such thing as a man who tran-scends the need of progress, whether intellectual or material. There is no human life that is "safe enough," "long enough," "knowledgeable enough," "affluent enough," or "enjoyable enough"—not if man's life is the *standard* of value.

. . . Freedom and production get blamed for the wrongs caused by government intervention—whether it is the side effects of "development" projects that would never have been undertaken except for organizations like the World Bank or the use of dangerous pesticides to replace safe ones that have been banned—as controls are added to overcome the problems created by earlier controls. Advocates of sustainable development claim to want the benefits that a free market provides, but they want to harness the power of the market—which means harnessing man's mind—to serve their own agenda, be it the poor or nature. . . .

The fundamental premises of sustainable development and of environmentalism must be challenged. Modern man is not destroying *his* environment; to the extent that he is free, he is constantly improving it. Value is objective, not intrinsic, and, for man, the standard of value is man's life. Production is a process of value creation by man, not of theft from nature. Government is the protector of individual rights, not the overseer of the Garden of Eden.

Human civilization is not courting environmental collapse—not, that is, unless it falls for the doctrine of sustainable development.

"*Environmental education efforts seem more focused on advancing particular political agendas than they are on educating children.*"

Environmental Education Indoctrinates Children in Environmental Extremism

Jonathan H. Adler

Not everyone favors the environmental curricula that have entered the public school system as an outgrowth of the environmental movement. In the following viewpoint, Jonathan H. Adler argues that what passes as "environmental education" is actually misinformation that glosses over an underlying political agenda. Adler also points out that the more important basic subjects of science and history have been displaced by environmental programs, leaving today's students with less understanding about the world they live in than previous generations. Adler is the associate director of environmental studies for the Competitive Enterprise Institute, a research organization that advocates private-sector solutions for environmental protection.

As you read, consider the following questions:

1. Identify two statements from the environmental curriculum that Adler labels as lies. What kind of information does he present to discredit them?
2. What does the author propose as an alternative to existing environmental education programs?

Excerpted from "A Child's Garden of Misinformation" by Jonathan H. Adler, *Consumers' Research*, September 1993. Reprinted with permission.

Environmental education has taken America's classrooms by storm. As of April 1992, 30 states had formal environmental education programs in place, and several more were pending, according to Environmental Education Associates. Further promoting the growth of environmental education are national associations and teachers' groups dedicated to expanding environmental awareness among American children. These include the Alliance for Environmental Education, the Community Environmental Council and the Global Rivers Environmental Education Network.

A Kid-Oriented Environmental Agenda

The federal government has gotten into the act as well, with the passage of the National Environmental Education Act of 1990 that established the Office of Environmental Education at the Environmental Protection Agency and created several national awards for green educational efforts.

Environmental messages aimed at children are not limited to the classroom. Television programs, comics and kids books are inundated with them. Children's heroes championing environmental subjects range from Captain Planet and the Planeteers and G.I. Joe to the Teenage Mutant Ninja Turtles and the Toxic Avenger.

These themes have spawned full-length feature films, such as *FernGully, The Last Rainforest*, and best-selling books, such as *50 Things Kids Can Do to Save the Earth*. The range of kid-oriented environmental media has led some commentators to note that Earth Day might as well be known as Kids' Day.

Yet despite the ubiquity of environmental messages aimed at kids, there is increasing evidence that children are not learning much of anything about the environment, save for simple platitudes and a blind faith in environmental causes.

One recent Roper poll found that American children are grossly misinformed about the environment. America's "Green Point Average," according to this poll, was only 31 percent, a failing grade by any measure.

Learning about the environment can certainly be a valuable educational experience. However, focus on the environment increasingly comes at the expense of basic instruction in important subjects, such as science and history. Without an adequate grounding in these disciplines, children will understand little about the world around them, let alone the environmental concerns that are now in vogue.

In many instances, environmental education efforts seem more focused on advancing particular political agendas than they are on educating children about the natural world.

Whether in the classroom or home watching television, children are bombarded by misleading and false statements about environmental issues. One frequent claim is that aerosol cans are capable of damaging the earth's ozone layer because they contain chlorofluorocarbons (CFCs). Although patently false—the United States actually banned CFCs from aerosols in 1978—this is asserted repeatedly in children's materials, including a 1993 textbook, *Earth Science*; a *Newsweek* "Just for Kids" environmental supplement distributed to schools nationwide; the Teenage Mutant Ninja Turtles book, *ABCs for a Better Planet*; the popular children's cartoon G.I. Joe; and Earth Day advertisements on Nickelodeon, a cable channel aimed at children.

It is no wonder that only 23 percent of children answered a question on this topic correctly in the Roper poll.

Environmental Automatons

As our nation continues its all-consuming pursuit of protecting the environment, we are overlooking the greatest cost of all: the toll on our children. . . . It is apparent, for example, that (1) children are being scared into becoming environmental activists, (2) there is widespread misinformation in materials aimed at children, (3) children are being taught *what* to think, rather than *how* to think, (4) children are taught that human beings are evil, (5) children are feeling helpless and pessimistic about their future on earth, and (6) environmental education is being used to undermine the simple joys of childhood. Are we raising critically thinking leaders or simple automatons that can recite the latest environmental dogma?

Jo Kwong, *The Freeman*, March 1995.

Unfortunately, this little green lie is hardly the only false statement presented to children. Consider this sampling of what's taught in leading textbooks and materials today:

• "Think of all the plastic items people throw away. None of them are biodegradable. Dumps and landfills are becoming filled with these items. We are running out of places to put them."—*Biology: An Everyday Experience.*

While most people believe that plastics are filling up landfills, paper is the most commonly landfilled item, accounting for more than 40 percent of the typical landfill's contents. All forms of plastics, including foams, packaging, and films, account for less than 20 percent of the typical landfill's volume. Moreover, while most plastic products are not biodegradable, little if anything will decompose when packed into a sanitary landfill. Excavations have found readable newspapers and half-eaten meals years after they were buried.

Given all of the benefits provided—including smaller and lighter packaging that reduces solid waste—the emphasis on eliminating the use of plastics is misguided.

Attacking Fossil Fuels

• "We must make a concerted effort to reduce our dependence on fossil fuels, such as oil, that create a lot of pollution when we use them for energy. Instead, we need to be developing and using more alternative fuels, like clean-burning natural gas. . . ." —"Saving the Earth" (A *Newsweek* "Just for Kids" supplement, 1993).

While there are pollution concerns raised by the use of fossil fuels, far less pollution is created per unit of energy produced today than in the past. Some environmentalists rail against gasoline, but ignore the fact that the advent of gasoline-powered vehicles has allowed for the conversion of thousands of acres of cropland into forest.

Moreover, advancing technology has dramatically reduced pollution from the burning of fossil fuels. It also must be noted that, contrary to the impression given by this passage, "clean-burning natural gas" is also a fossil fuel!

• "[D]uring the 1980s, automobiles and factories spewed deadly pollutants into the air daily. Fumes from American industries rose into the atmosphere and fell to Earth again as acid rain, damaging forests and lakes. Auto emissions contributed to clouds of smog hanging above the nation's largest cities. Smog-related illnesses and deaths increased sharply in the 1980s." —*American Odyssey: The United States in the Twentieth Century.*

As documented by the EPA, U.S. air quality improved dramatically in the past decade, and continues to do so. Many areas classified as "non-attainment areas" under the 1990 Clean Air Act would not be so classified today. Moreover, new cars rolling off the assembly line emit 95 percent less carbon monoxide and hydrocarbons than those produced 20 years ago. Last, but most important, there is no evidence to suggest that "smog-related illnesses and deaths increased sharply in the 1980s."

• "The supply of fossil fuels is being used up at an alarming rate. . . . Governments must help to save our fossil fuel supply by passing laws limiting their use."—*Biology: An Everyday Experience.*

Today, world proven reserves of oil and other fossil fuels are at an all-time high. While the 1970s brought concerns of imminent resource depletion, they were based on economic misunderstanding rather than empirical evidence. In fact, political controls are the sole source of concern over access to fossil fuel supplies today.

• "Perhaps the most frightening finding was the possibility of these chemicals [pesticides] ending up in the food we eat and the water we drink. Rachel Carson also had a serious concern over the possible connection between the widespread use of certain chemicals and the incidence of cancer in humans. . . .

"Her book made the best-seller list. Chemical companies fought back, spending thousands of dollars to disprove Rachel Carson's findings. They were not successful."—*Health: A Guide to Wellness.*

While many pesticides have been shown to cause cancer in animal tests, there is little evidence of any threat to the human population from pesticide residues on foods. As the pioneering work of Dr. Bruce Ames has demonstrated, virtually every "all natural" and "organic" product could be viewed as a health threat if held to the same standard.

While the writings of Rachel Carson effectively raised awareness about the impact of certain chemicals, such as that of DDT on bird populations, the majority of her conclusions about risks posed to humans have been disproved.

"Mis"-Educational Materials

Such erroneous presentations are in classroom materials for a variety of reasons. For one, teachers and the designers of educational materials are rarely well-versed in environmental issues and therefore accept assertions that sound plausible even though they lack scientific backing.

At an environmental education conference in June 1993, one well-received speaker presented claims about the impact of ozone depletion on livestock in southern Chile that have been rejected by scientists as well as by radical environmental groups, such as Greenpeace.

Unfortunately, few educators take the time to investigate these matters for themselves. Many classroom materials are developed by, or under the advisement of, leading environmental lobbying organizations.

The Sierra Club has been publishing environmental materials for children since 1977 and the World Resources Institute's teacher's guide is used in schools around the world. This results in materials that present the environmental view most conducive to these organizations' political agendas.

With a genuine lack of quality environmental educational materials, teachers have little choice but to use those offered by environmentalists if such issues are to be covered. If teachers aren't prepared to challenge or correct information that is wrong or mischaracterized, the inevitable result is that children learn an unhealthy dose of environmental fiction. . . .

Another element of environmental education is the intensifying effort to induce behavioral changes in children along environmental lines. This is one of the reasons for the promotion of environmental scares to children. Such efforts would always be of concern to parents, but these are particularly so given the reliance on misleading and inaccurate presentations of scientific information.

The effort to induce "correct" behavior has become a ubiquitous theme in green educational undertakings, from the best-selling *50 Simple Things* series to the "Kids C.A.R.E." cross-curricular environmental program.

Designed for students in grades 4-6, the teacher's summary of Kids C.A.R.E. activities lists a "student behavioral objective" for each activity. These range from simple things, such as inducing children to "politely refuse unnecessary bags for purchases," to "intercepting environmentally negative action."

The *Newsweek* "Just for Kids" supplement has its own list of environmental *Do's*—"wear sweaters inside when it's cold to keep the thermostat down," "at least 3 people in a car," "air-dry laundry," and "buy rain-forest nuts"—and *Don't's*—"buy products in plastic bottles," "load groceries into doubled plastic or paper bags," "buy teak or mahogany wood," and "leave broken windows unrepaired."

A "student resource book" produced by the Close Up Foundation [a civic organization promoting political involvement of citizens of all ages] charges that "perhaps our most disturbing contribution to the environmental problem is our 'throwaway' attitude." This attitude, according to the booklet, has given Mother Earth a "'cancer' growing inside her. . . . It is a cancer of trash and waste."

Unfortunately, this encouragement of "ecofriendly" attitudes often leads to environmental dictates that do little to protect the environment. The condemnation of plastic is a case in point.

Plastics provide many important benefits, not the least of which is reducing the weight and volume of packaging, thereby lessening the flow of wastes to landfills. In addition, plastic products are often less expensive than those made from other materials because they require less energy, water, and other natural resources to produce and transport. Even a third-grader should be able to recognize the environmental benefits of that.

Saving the Earth

An increasing number of materials explicitly encourage political action. Children are told to write letters to Congress, draft petitions and boycott products that are not deemed environmentally correct. In at least one instance, elementary school children even filed comments with a federal agency about a proposed regulation of a highly technical nature. *Health: A Guide to Wellness* provides an "improving yourself" section that tests students' "efforts to effect change."

In this test, children receive points by agreeing to write letters to government officials and petitioning against residential development. Child heroes, such as the Teenage Mutant Ninja Turtles, are repeatedly telling children to "write to your government at every level—city, county, state, and federal."

A "Save the Earth Action Pack," distributed to schools in 1992 by the Turner Broadcasting System, even went so far as to tell children "to increase the amount of time and money" given to environmental lobbying organizations.

A good example of the growing move toward attitude alteration is the environmental education plan recently developed by the New Jersey Environmental Education Commission. Based upon guidelines from a United Nations environmental education conference in 1977, this program is quite explicit in its intent to "develop in citizens of the state the knowledge, attitudes, values, skills, and behaviors needed to maintain, protect and improve the environment."

What are these "values"? Well, they include the fact that "environmental issues have a moral and spiritual dimension" and that the "diversity of culture" should be considered in environmental policies. Moreover, the commission "believes that all New Jerseyans should be able to . . . develop a lifestyle that promotes environmental awareness." At its worst, as in this case, environmental education has begun to approach indoctrination.

Unfortunately, many of the current problems with environmental education are caused by a genuine lack of quality classroom materials that deal with environmental issues. Rectifying the existing problems therefore requires more than heightened sensitivity in the selection of textbooks and curricular activities.

Concerned parents and educators have little choice but to balance misleading materials with independent research into the scientific and economic facts surrounding specific issues in order to ensure that the environment is covered in a fact-based and non-ideological manner.

Certainly environmental issues can be a valuable addition to any child's education. Coverage of issues, such as the nature of ecosystems, hydrology, and climate patterns, can enrich educational experiences. These issues are more than appropriate for inclusion in conventional science curricula, and the passage of significant environmental statutes should be covered in history classes.

As children get older, basic instruction in scientific matters can give way to discussions exploring the pros and cons of various issues. Some schools have, in this vein, taken to setting up classroom debates about issues such as recycling and policies to address global warming. This can be a valuable learning experience.

Teaching children to understand these issues may not always produce the desired point of view, but it will make them more eco-smart.

"Conservative groups have . . . accused schools . . . of threatening their communities' economic foundations by teaching students . . . the effects of industrial pollution."

Environmental Education Teaches Children About the Environment

Barbara Ruben

Industry, in concert with the religious right, is mounting a campaign to restrict environmental education in the public schools, according to Barbara Ruben. In the following viewpoint, she argues that environmental curricula are being unfairly labeled "anti-family or anti-Christian," and that pressure from parent groups and industry has school administrators and teachers afraid to address basic environmental issues in the classroom. As a result, Ruben contends, environmental education is being severely compromised, depriving students of the freedom to learn about important ecological realities. Ruben is the editor for *Environmental Action* magazine.

As you read, consider the following questions:

1. Name two restrictions reported by Ruben that the Meridian, Idaho, school district has enacted in its classrooms.
2. According to this viewpoint, what group is affiliated with the religious right, and who funds it?
3. What questionable assertions, according to Ruben, does environmental policy analyst Jonathan Adler make?

Barbara Ruben, "Reading and Writing, but Not Recycling," *Environmental Action*, Spring 1994. Reprinted with permission.

In suburban Boise, an elementary school music teacher altered a skit performed by students because she feared it was too controversial to pass muster with her conservative school board. The song she removed from the skit had nothing to do with the usual gamut of sticky issues grappled with by the schools such as AIDS or sex education. Rather, it taught about the virtues of recycling.

In Laytonville, California, parents campaigned to kick Dr. Seuss's *The Lorax*, who "speaks for the trees," off the second-grade reading list for being anti-logging in a town where the major industry is timber.

And in Broken Arrow, Oklahoma, a fundamentalist Christian group attempted to remove a teacher resource manual on the environment, called *Earth Child*, from elementary schools. According to the group Oklahomans for Quality Education, the book promoted, among other things, satanism, eastern mysticism and worship of the Earth.

Excising Environmental Education

The religious right and conservative groups have taken up the cause of excising or severely curtailing environmental education in the classroom. They've accused schools of teaching satanism, of lying to children about the threat of global warming and ozone depletion and of threatening their communities' economic foundations by teaching students to critically evaluate the effects of industrial pollution.

At the same time, a variety of conservative magazines—from the libertarian *Reason* to the Unification Church–owned *Insight*—are working to incite action against environmental curricula. Writes reporter William Grigg in the bi-weekly *New American*, "Environmentalism is largely based on a kind of secular religion that seeks temporal salvation through an all-powerful global government. . . . The growing 'environmental education' movement is a recruitment drive intended to conscript young students into a pagan children's campaign."

And jittery school boards, worried about the wrath of parents and conservative education critics, have in some cases transformed the way kids learn about the environment.

In the Meridian, Idaho, school district, for example, 1993 teaching guidelines issued by the school board specify: "Discussion should not reflect negative attitudes against business or industry who do the best job under present regulations considering economic realities." This edict frightened a teacher into pulling the song on recycling from a school skit and caused another to face an angry principal after teaching about reintroduction of wolves into Yellowstone National Park. The same district required parents to sign permission slips before a fourth-grade class could discuss the presidential election, and a teacher removed an ACLU [American Civil Liberties Union] poster about the Bill of Rights from her classroom after a parent objected to it.

Under Attack

"I'm outraged and I'm angry. The students' education is being short-changed," says fourth-grade Meridian teacher Lee McGlinsky. McGlinsky says she has been careful to balance environmental lessons in her classroom, but she's not going to stop her students from a recycling project or learning about endangered species. Even with these seemingly innocuous subjects, McGlinsky says, "I know I'm treading on touchy ground." What gives McGlinsky courage to boldly move into this tricky territory? She's retiring after this year, worn down, she says, by the increasingly restrictive climate that hampers her teaching.

Two years earlier in Bend, Oregon, the school superintendent canceled a four-day environmental camp for students in grades six through eight called Earthkeepers. Bowing to pressure of parents who objected to the program's spider-web symbol as "anti-Christian" and song lyrics such as "hello sun" as "paganistic," the cancelation came just days before the camp was scheduled to begin, causing the schools to lose $14,000 in non-refundable fees for the camp. The local logging industry also asked that the camp program be scrapped because of a tree hugging exercise and other lessons it felt did not portray "all sides" of environmental issues.

Across the country, other teachers are also facing increasing attacks on what and how they teach. The liberal research group People for the American Way each year documents hundreds of such incidents in a report called "Attacks on the Freedom to Learn." In the 1992–93 school year, the group uncovered 395 examples of attempted restrictions on books or curricula in all subjects, the highest number in the 11 years the group has done the report. In 41 percent of the cases, objectors succeeded in removing the materials from schools.

Who Is Leading the Offensive?

"The criticisms of environmental education are part of an overall attack on schools by the religious right—about 35 percent of the attacks we document—as what they see as anti-family or anti-Christian," says Deanna Duby, director of education policy for People for the American Way. "The environment is attacked most in communities where industry—particularly extraction of natural resources—is involved. The wise use movement is closely linked to the religious right and so there is an overlap in some of the attacks they've launched." Wise use groups, backed heavily by industry funding, have increasingly opposed environmental legislation in many western states.

Leading the offensive for the religious right is the California-based group Citizens for Excellence in Education (CEE), which claims to have 1,200 chapters that have helped elect 3,500 fundamentalist Christians to school boards across the country. In his monthly letters to members, CEE president Robert Simonds' rantings range from rage against teachers "using our school classrooms for hypnotic exercises on oneness with 'mother earth,'" to his Earth Day message in 1992 where he said, "Satan uses the evil in the occult new age witchcraft lessons in our classrooms to divert our children's faith away from the true and loving God toward the new age god of 'Mother Earth' while our school teachers and administrators are saying, 'Well, it's good environmental ecology.'"

The Mission

A fifth grader places her hands on her stomach, and runs toward her classmates. She's playing "Oh, Deer!", pretending to be a hungry doe. If she tags a classmate making the food sign, she lives; otherwise she dies. "Oh, Deer!" is the favorite activity of the very popular *Project Wild*.

When describing *Project Wild*'s goal, director Dr. Cheryl Charles could just as well be defining environmental education's mission. "It's to assist learners in developing awareness, knowledge, skills and commitment that result in informed decisions, responsible behavior and constructive action concerning wildlife and the environment upon which we all depend."

Dr. Woodward Bousquet chairs environmental studies at Warren Wilson College in Swannanoa, North Carolina. He says "the house of environmental education is built upon a three part foundation: natural sciences, social sciences and citizenship skills. People need firm understandings of ecology, economics, politics and ethics to be able to make environmentally-responsible decisions."

Mike Weilbacher, *E Magazine*, March/April 1991.

CEE has been instrumental in rallying against the book *Earth Child*—which includes simple exercises to introduce young children to the beauty of nature and to learn about the stars—calling it satanic and claiming it uses "subliminal messages to brainwash our children."

Despite a well-orchestrated campaign by CEE, the school board in Broken Arrow, Oklahoma, voted to retain the book in district schools. However, just days before the board's decision,

CEE did pressure a local utility, Oklahoma Public Service, to withdraw its sponsorship of teacher workshops on the book.

Although CEE officials declined to be interviewed for this viewpoint, Focus on the Family, a more moderate group advocating teaching of Christian values in schools, agreed to share its perspectives. Focus on the Family has no qualms about teaching kids to recycle or conserve energy, says Tom Hess, the group's magazine editor. But "it's very difficult for a pro-life Christian to find an environmental education program that isn't zealously advocating abortion as a way to stop pollution," Hess says.

Questionable Science—or Assertions?

Although employees of Focus on the Family do not lobby schools to remove materials the group finds objectionable, its magazine instructs parents to "encourage your local school to teach environmental studies, but without abortion advocacy or mystical emphasis."

Hess also criticizes Captain Planet, a cartoon whose eponymous main character fights such polluting villains as Looten Plunder and Duke Nukem. Imbued with help from Gaia the Earth spirit, cartoon children help Captain Planet protect the Earth. Hess says he has warned parents about the cartoon's "New Age and population control themes" through the group's magazine.

Captain Planet and other children's cartoons like the Teenage Mutant Ninja Turtles have also come under fire by right-wing critics. These "simple-minded and inaccurate" depictions of environmental issues are scaring children and giving them incorrect information, says Jonathan Adler, an environmental policy analyst with the Competitiveness Enterprise Institute in Washington, D.C., and outspoken environmental education critic. "In many cases, kids are being taught science at school that is questionable, one-sided or patently false," he says.

But in articles for the conservative think-tank the Heritage Foundation, the *Wall Street Journal* and *Consumers Research Magazine*, Adler makes his own very questionable assertions. In an article titled "Little Green Lies," he says that, although recycled paper can be used for newsprint and cardboard boxes, other uses of recycled paper are inappropriate because of its inferior strength. Adler also asserts, "Fear of the risks of pesticide residues on food are greatly overblown" and claims that a Teenage Mutant Ninja Turtle book that says "acid rain pollutes rivers and kills fish and trees" is untrue. Adler goes so far as to claim that acid rain actually helps eastern forests by providing the nutrient nitrogen.

Adler also takes environmental education curricula to task for promoting activism. "Schools should not encourage political advocacy. Teachers should not be telling students to sign petitions or write letters to their legislators. That's not education," says Adler.

Replacing "Objectionable" Learning Materials

To combat what he calls "political indoctrination" of children, Adler and several others are now putting the finishing touches on a book called *A Parents' Primer on the Environment*. The book contains a critique of textbooks that talk about environmental issues and questions such issues as global warming and ozone depletion.

"Ninety-five percent of textbooks only present the scientific evidence that global warming is occurring. They ignore the two other alternatives: that global warming isn't occurring at all or that warming may occur, but it will result in benefits for the environment," says co-author Michael Sanera, who is president of the Arizona Institute for Public Policy Research. "Textbooks grossly ignore scientific evidence to the contrary on a variety of issues. I believe they're doing it deliberately to get kids politically active," Sanera blusters.

A Parent's Primer is also directed at teachers, Sanera says. "We can see that many teachers will find it very valuable. Teachers who want to round out their textbooks will be able to add lessons looking at why in the paper cup versus Styrofoam debate, paper is no more environmentally correct.". . .

Citizens for a Constructive Tomorrow (CFACT), an organization that "promotes free-market and safe technological solutions to current consumer and environmental concerns" and is funded heavily by industry, tested its anti-environmentalism educational materials in southern California high schools. The group hopes to take its program nationwide. CFACT's program "discusses the immeasurable benefits to our daily lives of coal, oil and nuclear power" and covers "the safety and importance of burying and burning our trash."

"There is an intentional effort under way in our schools to indoctrinate children with a dangerous Green mentality. . . . If the Greens already have a foothold in your local schools, you should bring it to the attention of your friends and neighbors and seek to remove this dangerous curriculum," writes CFACT President David Rothbard in an appeal to members asking them to fund CFACT's education program. In the group's newsletter, he describes environmental education curricula currently in schools as "a lavish buffet of pop science, end-of-the-world hysteria and touchy-feely New Age gobbledygook."

Another curriculum with a right-wing slant is called "Visions of the Future." Developed by New York University humanities professor Herbert London in 1986, the teacher guidebook and social studies textbook "eliminate the element of environmental extremism found in many textbooks and offer a cogent and thoughtful exploration of environmental issues, especially cost-benefit analysis," London says. The books have been used in New York, Arizona, Indiana and North Carolina.

A Tiny Fraction

At the same time as anti-environmental programs are gaining a higher profile, materials from environmental groups are still in high demand. However, there is increased competition from industries to tell their "side" of the story, says Midge Smith, who directs educational programs for the National Audubon Society. Smith says she knows of no cases where Audubon's or other environmental groups' materials have been removed from the classroom, but she acknowledges that "it would be tough trying to market something like forest education materials in timber country."

Although restrictions on environmental education have made a palpable impact on schools, People for the American Way's Duby remains optimistic. "Overwhelmingly people are opposed to censorship. Very often it is only a tiny fraction—although a vocal one—of a community that wants to restrict education. Parents need to speak out, they need to keep up with what's going on in the schools, and as taxpayers they have an obligation to keep education from being fettered."

"*Conservation is a public good. Government coercion is a political evil. And the two things cannot go together.*"

Environmental Regulations Should Be Curtailed

Fife Symington

The federal government, goaded by environmentalists, has overstepped its bounds in regulating for protection of the environment, Fife Symington argues in the following viewpoint. Government agencies are enforcing unreasonable and impractical requirements, he maintains, and are trampling the freedoms essential to democratic society in the process. Natural resources should be managed by those who live near them and use them, he contends, not by bureaucrats in Washington who are ignorant of unique regional conditions. Symington is the governor of the state of Arizona. This viewpoint is adapted from a speech he delivered to the Heritage Foundation on June 19, 1995.

As you read, consider the following questions:

1. How does Symington define "environmentalism"?
2. What does the author name as the two problems that Americans must overcome if they are to be effective stewards of nature?
3. What, according to Symington, is missing from the environmental movement?

From "A Federalist's Approach to Protecting the Environment" by Fife Symington, *Heritage Lectures*, no. 534, 1995; ©1995 by The Heritage Foundation. Reprinted with permission.

185

Whenever I join the environmental debate, as I often do, I'm reminded of a problem that conservatives always face. The left regard our environment as their issue, their ideological property. Whatever ideas we might have, whatever arguments we might offer, our motives are always suspect. Here as elsewhere, liberals have a way of turning every rational debate into a contest of emotional authenticity. Any who would challenge them must first demonstrate enough "sensitivity" for the liberals' satisfaction.

The absurdity of this posture strikes me every time I come to Washington. Often I'm here seeking relief for my state from one or another environmental regulation, edict, or fine. This means constant haggling with Washingtonians eager to tell me what's best for Arizona.

It's always fascinating to leave behind my hiking boots and the mountains of Arizona, and come here to be lectured by "naturalists" whose rugged trails run through Georgetown, Dupont Circle, and Rock Creek Park. In fact they seem to revere everything about our state—except for the opinions and livelihoods of the people who live there.

Some Basic Common Sense

I have been asked to describe a federalist approach to protecting the environment. Most environmentalists, I suppose, would regard that juxtaposition as some sort of bold intellectual synthesis, as if the two ideas were opposites to be reconciled. By philosophy and instinct, they just equate the environmental cause with centralized federal power. I'll explain why I reject that connection—why, in fact, just the opposite is true. I'd also like to set down some specific proposals for reform and conclude with my vision of wise environmental reform.

Let's begin in the most basic, common-sense terms.

Federalism is not some elaborate theory of government. It's the simple insight that most problems are best left to the people nearest to the problem, people with direct knowledge of the circumstances. The more you remove the influence of local power, the more you atrophy the dynamism of local decision making. Government in a federalist system tends to be more efficient, and the people tend to be freer.

Environmentalism, stripped of the quasi-religious nonsense that today often goes with it, amounts in the end to simple stewardship. At its best, it's the sense all decent people have that with the natural riches given humanity comes a duty to use them wisely. A moral duty, a civic duty, but a quite practical one as well.

To me, the connection between these ideas—stewardship and self-government—could not be more apparent. Obviously, we have to manage our land, water, air, timber, minerals, and wildlife with care. And just as obviously, that duty is usually best understood and carried out by the people living upon that land.

Sure, there are exceptions. We all recognize that some resources and natural treasures are a national responsibility. Historically, I suppose you could trace this sense of national responsibility back to 1849, when we established the Department of the Interior. In fact, we should restore the Department to its staffing level of 1849. It was the first federal department created after 1789, so its federalist credentials are pretty sound. There is a consensus as to our national stake in good stewardship, and it is of long standing.

But if we have this consensus, why is there so much bitterness to our environmental disputes? As I see it, there are two reasons.

First, the bitterness, the anger, the endless controversy arise from the methods by which environmentalists and the federal government attempt to carry out their aims. Good stewardship, it seems to me, has to begin with a higher opinion of human nature than many environmentalists today bring to the matter. There would be a lot less bitterness if they did not rely so much upon government dictate and coercion, if they were not always eyeing their fellow citizens with such deep distrust.

My brand of environmentalism begins with these three assumptions: Conservation is a public good. Government coercion is a political evil. And the two things cannot go together.

The other, related reason is that many environmentalists today have forgotten Theodore Roosevelt's reminder that "conservation means development as much as it does protection." With Teddy Roosevelt, I reject the notion that economic development and environmental stewardship are natural, predestined enemies. In fact, from modern history I draw exactly the opposite lesson.

A Complete Disaster

Think back for a moment to our first post–Cold War glimpse of the dead lakes and dying forests of the old communist world. This was stewardship, socialist style. And yet even today, it's hard to shake the suspicion that for some environmentalists, the cause is just a pretext for airing their larger grievances against the free market. The extreme wing of the environmental movement has been well described as America's last enclave of socialism. For them, every new industry seems to pose a menacing ecological threat.

It is the free market that is moving us away from the old smokestack industries and into the age of the computer, which does not pollute and is now our prime aid in understanding and safeguarding the environment. In fact, to free enterprise we owe all the technologies that today make for a more efficient use of natural resources. Think of just about anything in modern life which affords ordinary people comfort, health, peace, and security—and you'll find a product of free enterprise. What's more, in free market economies alone do people even have the wealth, education, and luxury to sit back to reflect on our duties to nature. Put another way, how's the environmental movement shaping up in India or Cuba?

My own belief is that if we can overcome these two problems—the coercion of government and the snobbery of environmentalists—Americans can truly come together to meet the challenges of nature. This is just another way of saying that the best environmental policy is democracy.

Now, let me give you some examples of the kind of problems we've run into in Arizona. We had one particularly interesting incident in early June 1995.

Back in the previous fall, an environmental activist group went to federal court seeking an order that the Fish and Wildlife Service come up with critical habitat designations for the Mexican spotted owl. The judge was Carl Muecke, a retired Federal District Court judge. He duly issued the order. But there were a couple of problems. First, the plan Fish and Wildlife came up with, about four million acres altogether, included Arizona's Prescott National Forest—where, literally, no owl had ever been spotted. And second, between court order and deadline, Congress passed a moratorium on new federal regulations. Faced with this, the good judge replied in essence: "Do it anyway!"

Pushing Back

So, against the will of Congress, the plan went into effect. Result: All the state's own forest health initiatives must be put off. We have gridlock in our national forests. In the end we will destroy these habitats. And Mexican spotted owls now have a new vacation spot—only they don't read court orders, and so they still don't come anywhere near Prescott National Forest. Congressional will was simply disregarded by one arrogant federal judge goaded on by a litigious band of environmental zealots.

Someone has observed that deep down in every liberal is a commissar yearning to be saluted. On our federal benches are the finest examples of the trait. And with the electoral tide turning against liberals, the situation may get a lot worse before it gets better.

Here's another example. I have been urging Congress, and will again at every opportunity, to repeal the Endangered Species Act. Why? Because it is a complete disaster—to the states, the people, and even the species it was meant to help. Because it has thrown entire regions of the Western United States into profound uncertainty. Because, however noble its original aim, in practice the law has become a tool for radical environmental groups to shut down entire industries. Even more bewildering, it prevents the wise management by the states of forest lands, which has led to pestilence and fire.

The Environmental Protection Agency (EPA) has long been pushing Arizona to adopt what's known as the California standard of air quality. We've been pushing back, so the case is still in litigation.

In all its philosophical complexity, our basic argument is this: Arizona is not California. For example, we have a lot more desert land than they do. The desert raises up a lot of dust and various pollen from its unique vegetation. As a matter of fact, we have found that if the California standard were forced upon Arizona, we could not meet it even square in the middle of Organ Pipe National Monument, which is square in the middle of nowhere.

We've also pointed out to the EPA that air quality is an inconstant thing. When there's construction underway in a given area, obviously that's going to affect the air quality—but only temporarily. So, for example, the EPA will take its measure in Chandler, where we currently have thousands of new homes under construction and the largest industrial project on Earth.

I could tell you many more such stories—I have a collection of them I'll match with any governor in the Union. Solomon Bridge near Safford, Arizona, was washed out, requiring people to drive their kids to school on dangerous roads. The Fish and Wildlife Service said that they couldn't fix the bridge because of an endangered minnow that lived in the river beneath the bridge. We told the people of Safford to go ahead and fix it anyway. They went into the river and fixed the problem, and nobody has heard a peep from the Fish and Wildlife Service.

Restoring Self-Government

But let me end by trying to sort out the lessons I draw from these experiences. Given the audacity of environmentalists, and their cavalier approach to democratic procedures, how can we restore self-government to environmental decisions?

I believe we need a constitutional amendment imposing term limits on federal judges: Ten years, and it's time for a good long hike into the real world. They need to go back and live in the world they helped create. This would put an end to the scandalous abuses of our courts by environmental groups, among many other liberal activist groups that can advance their aims only by judicial dictate.

That's reform number one. If any of you doubts this can be done, I would remind you that there is a political revolution going on. Even if you don't hear it talked about in the U.S. Senate, people are fed up. We are going to return power to the states. States are getting ready to receive this power. And we didn't get this far by thinking small.

My second proposal is purely organizational. We talk loosely of "federal environmental policy." But, really, there is no single coherent policy to speak of. What we actually have is a confusion of federal voices, each barking a different set of orders at each state and often competing for bureaucratic turf against one another. In land management the problem is particularly costly and inefficient.

©Boileau/Rothco. Used by permission.

We have read threats of closing national parks and the reinvention of the Forest Service. Unfortunately, in both instances the President has offered sophomoric solutions to significant management problems. The White House and their agencies need to get out of the box and look at efficient reorganization. They must accept the challenge of smaller budgets as an opportunity to make public lands more beneficial to all Americans. I will offer Congress a deal: You give our State Parks Department management authority over three of your federal monuments and 90 percent of current funding, and we will run them more efficiently and, more than likely, make a profit.

The situation today is intolerable. There is too much overlap, too much rivalry, too much confusion—and too little actual service to the public.

I have a third proposal: The President should also streamline the Environmental Protection Agency. There are approximately 8,000 EPA employees in the regional offices. It is unclear what these employees do that is not already done by the states. There is nothing that is done in the regional offices that cannot be done in the state environmental agencies.

I recognize and agree that there needs to be a functional liaison between the states and EPA, but I also believe the present EPA organization is an overly expensive and an increasingly ineffective way to interact with the states. I propose we eliminate the regional offices and establish state offices that are co-located with the primary state agency responsible for the implementation of the nation's environmental laws. Each state office would have 25 to 50 federal employees to work with the state to implement federal laws.

This partnership would allow for programs to be developed once in partnership rather than developed by each state, only to be second-guessed by the regional offices. This arrangement would also allow the federal employees to better understand state issues because they would be directly involved from the beginning. Even if each office had 50 employees, this would only be 2,500 nationwide—a reduction of over 68 percent from current staffing.

Scrap the Whole Thing

Next, there is the problem of unfunded mandates. Congress, of course, has done something about this. But, frankly, the central problem remains. It seems almost too obvious to point out, but the staggering thing about these mandates is the undemocratic capriciousness of it all. Hardly anyone even dares question the practice. Seldom are we even dealing with specific provisions in the laws Congress has passed.

Over the years, Congress fell into the habit of enacting grand new laws as confusing in effect as they are high-sounding in name: the Clean Air Act, the Critical Endangered Species Act, whatever. And then, well, they just left it for the bureaucrats to sort out the details. Not surprisingly, the details invariably turn out to confer greater and greater powers upon the interpreters. Challenged by citizens thrown out of work by their regulations, or in danger of losing their property or seeing its value plummet, these various agencies then seek out like-minded jurists like Judge Muecke. And so, using yet more public money to litigate, they almost always get their way. It is truly an insidious practice. The freedom revolution will not be complete until we put a stop to it once and for all.

In the long term, we need to scrap the whole irrational regulatory structure that has been dragging our economy down, costing us tens of billions of dollars, and ensnaring countless productive citizens trying to make a decent life for themselves.

There are tens of thousands of people in Washington whose sole mission in life is to think up and enforce environmental regulations. Between these rule-makers and the ruled, there is a chasm wider than any in the natural world. The folks making policy are too far removed from the folks living under it. If we are serious about reform, we will retain only a very few of these rule-makers, placing statutory limitations on the number of people each agency may employ. What will become of all the others? Phase them out, buy them out, do whatever it takes. But it is time for them to pack up their knapsacks. Somewhere down the trail lies an honest living.

That's the long term. In the short term, we should apply a sunset law to all existing federal mandates. After a given period—say, five years—all regulations that are not demonstrably protecting the health and safety of the American people should be scrapped. . . .

We are critical of the environmental movement of today. But at the same time, there is much good within it. It comes from a very noble impulse. Within it are many sincere and idealistic people. We need such people.

But at least in its political ranks, I often sense that something is missing from the cause. There is something fearful, grudging, resentful, and deeply pessimistic about environmentalism as we hear it advanced. They forget about human ingenuity, the ability of free people to rise to the problem, to produce more, to be good stewards of the Earth. They forget human nature, which is never quite as blind and grasping and brutal as our more fretful environmentalists seem to assume.

I guess I prefer the term "stewardship" because it captures both the responsibilities and the opportunities we meet every time we step into nature's province. It captures both sides of human nature—that side which bows before nature and that side which rises to master it, rises to build homes and cities and lay up stores while still leaving those spaces of quiet in between. I like it, too, because it conveys a little bit more humility than we observe among today's environmentalists. This is the sense of the old saying that God creates and people merely rearrange; that "Nature, to be commanded, must be obeyed."

My friends, these are revolutionary times. On this issue, as elsewhere, the moment has arrived. In place of fear and fretting and false alarms, we must offer a vision of true stewardship—reclaiming our nation in every sense. It is not a time to be timid. The moment has come to reclaim, all at once, the rich endowment entrusted to America by nature and the freedom entrusted to us by the Founding Fathers.

"What remains to be seen is whether revamping environmental policy will [jeopardize] . . . the great strides made over a quarter-century."

Curtailing Environmental Regulations May Prove Harmful

Frank Clifford

The election of November 1994 marked an important victory for the Republicans, who won a majority status in both congressional houses for the first time since 1952. Environmental groups, as a result, faced a dramatic shift from the Democrat-controlled Congress they had lobbied successfully in the past. The newly elected Congress announced plans for a complete overhaul of environmental legislation. While conceding that federal regulations have not always been reasonable in terms of cost or efficiency, Frank Clifford argues in the following viewpoint that the drastic cuts proposed by the 104th Congress could prove devastating for the environment. Clifford is an environmental writer for the *Los Angeles Times*.

As you read, consider the following questions:

1. How have environmental conditions changed since the first Earth Day, as Clifford sees it?
2. Name two intentions Clifford attributes to Republican efforts to roll back environmental regulations.
3. Why does the EPA's Carol Browner label proposed cost-benefit regulations "absurd," according to the author?

On the eve of Washington festivities honoring the 25th anniversary of Earth Day, Louisiana congressman Billy Tauzin told a story about dynamiting fish in a lake.

It was a funny story full of the self-deprecating Cajun humor that makes him a beguiling speaker. But when Tauzin, a Democrat who is a leader in the movement to rewrite the nation's environmental laws, got to the punch line, he wasn't joking anymore.

"It's time now to flip that dynamite right into the water," he told a roomful of shopping center developers who had gathered for a daylong skewering of laws protecting wetlands and endangered species.

A Lit Fuse

Some would say that the fuse has already been lit, and that it's only a question of how big the explosion will be as Congress—in the name of common sense, cost consciousness and private property rights—moves to overhaul the laws enacted since Earth Day 25 years ago today [April 22, 1995] marked the beginning of the modern era of environmentalism.

Targeting landmark laws on air and water quality as well as the preservation of wilderness and wild animals, the campaign amounts to a counterrevolution against the federal government's license to regulate in the public interest.

The reformers, mostly Republicans, challenge the very assumptions underlying a generation of environmental policy. They say too many expensive regulations are based on unscientific judgments and that billions of dollars have been spent to regulate activities that pose insignificant threats.

John D. Graham, a professor of public policy at Harvard's School of Public Health, may have stated the argument as succinctly as anyone when he told Congress recently: "We regulate some nonexistent risks too much and ignore larger, documented risks. We suffer from a syndrome of being paranoid and neglectful at the same time."

A thousand-page law, the 1990 Clean Air Act, requires the expenditure of billions of dollars "to clean up the last 10 percent or so of pollutants in outdoor air," Graham has written, while comparatively little attention is directed at improving the quality of air indoors "where people spend more of their time."

What remains to be seen is whether revamping environmental policy will eliminate the inconsistencies and excesses of the Washington bureaucracy without jeopardizing the great strides made over a quarter-century.

"What I have been seeing over the last 100 days is a frontal assault . . . an attempt to roll back 25 years of public health and environmental protections," said Carol Browner, the Clinton Administration's chief of the U.S. Environmental

Protection Agency (EPA).

Since the first Earth Day, air quality in much of the country has improved substantially. Airborne lead has all but disappeared since leaded gasoline was phased out. Smog levels have been cut by close to one-half in the Los Angeles area despite a large increase in automobiles.

Nearly double the number of lakes and rivers are swimmable and fishable since the passage of the Clean Water Act in 1972. Untreated sewage is no longer dumped in waterways. Recycling has led to a 20 percent reduction in municipal waste buried in landfills.

In the same quarter-century, the amount of federal land set aside for wilderness or other conservation purposes has jumped from about 50 million acres to about 280 million. National park land has nearly tripled, to about 80 million acres.

Environmental groups say that the job is far from finished, that the nation needs to continue to fix the damage done by industrialization and careless growth while preserving a natural heritage for future generations.

But driving the reform movement is the belief that many of the old solutions have backfired.

The revisionists argue that pesticide regulations have caused farmers to grow crop strains that are high in naturally occurring poisons, that logging prohibitions have clogged national forests with dead and dying timber, providing the kindling for disastrous fires, and that hunting bans in national parks have upset the predator balance and led to huge herds of elk, deer and bison that spread disease and devastate vegetation.

More than anything, the critics say, the way the environment has been protected in the past 25 years has burdened Americans with unnecessary costs, put too much valuable land off limits to commercial development and ordered up remedies that were not based on sound science.

"We can no longer afford poorly targeted, inefficient regulations that achieve only marginal environmental benefits in an inflexible manner and at an excessive cost," said Jerry Jasinowski, president of the National Association of Manufacturers, one of the driving forces behind the movement to change environmental laws.

The Momentum Refuses to Slow

The Clinton Administration has tried to slow the reform steamroller with concessions, agreeing, as Browner said in an interview, that "the process needs to be changed, that we need to develop more innovative, cost-effective solutions to environmental problems."

The Administration has offered industries more incentives to come up with their own solutions to pollution problems. And it has agreed to exempt owners of small tracts of land from regulations under the Clean Water Act and the Endangered Species Act.

But these gestures have done little to slow the momentum for change. [In 1995] there appear to be enough votes to make substantial revisions in the Clean Water Act, the Endangered Species Act, the Safe Drinking Water Act and the Superfund law governing cleanups of the nation's most hazardous toxic waste sites. In most cases, pending legislation would sharply limit the federal government's enforcement powers.

Signe Wilkinson/Cartoonists & Writers Syndicate. Reprinted by permission.

Even before changing those laws, this Congress has demonstrated an ability to work around environmental restrictions. Determined to allow large increases in logging in national forests, Congress headed off anticipated environmental challenges by declaring in advance that all forest protection laws had been complied with.

Republican leaders of key committees in the House and Senate have also made clear their intentions to reduce by millions of acres the amount of land set aside as wilderness and to turn some of that land over to states or private owners. Senior members have called for opening up Alaska's Arctic National Wildlife Refuge to oil drilling, paring down the acreage in the national park system and allowing more commercial activity—such as cruise ships and tour buses—in certain parks.

Shell-shocked environmentalists say the congressional agenda goes far beyond any public clamor for reform. They point out that few 1994 election campaigns focused on environmental issues, and they quote opinion polls that indicate continued public support for environmental protection.

"There's no public mandate for any of this," said Carl Pope of the Sierra Club. "That's what is so outrageous about it."

But others are not so sure. Former Oklahoma representative Mike Synar, a Democrat who chaired the Environment, Energy and Natural Resources Committee until his 1994 political defeat, says Democrats should have seen the storm coming.

"The polls said that the public cared about the environment, but the expressions on people's faces told you something else," Synar said. "People were tired of having the government in their face. Tired of gun bans and smoking bans. The American public wants a good letting alone."

Synar says the momentum for change would not be what it is if Congress under Democratic control had done more to curb examples of what he called regulatory silliness.

"Take the Safe Drinking Water Act," Synar said. "There was no reason to make some hard-pressed municipality test for 150 different contaminants that hadn't shown up in a thousand miles of the place. But we held our ground stubbornly. We said you gotta test."

Populist Tactics

It is no accident that the forces pushing for an overhaul focus less on the environment than on the issue of government meddling. In his remarks to the developers, Tauzin said the drive to change environmental laws was doomed as long as it was seen as a campaign against spotted owls, timber wolves or sea turtles.

But if it were a movement to save a Northern California sawmill, a Montana rancher or a fourth-generation Louisiana fisherman from the snares of an environmental bureaucracy, Tauzin said, then it might succeed.

And that is the way the campaign is being presented—as a populist movement to vindicate the property rights and financial security of working-class Americans.

Congressional hearing rooms reverberate with stories of excess—of a home condemned because it was built unknowingly on a wetland, of an immigrant farmer threatened with prosecution for mowing a field where an endangered rodent lived, of a fisherman who hung himself because he could not make a living as a result of environmental restrictions on the nets he could use.

These stories, however, are not always what they seem.

Tauzin talks about a 14-year-old Boy Scout lost in New Mexico's Pecos Wilderness and forced to spend an extra 24 hours in the mountains after he was spotted by a helicopter. Tauzin blames the delay on the reluctance of U.S. Forest Service officials to violate a "crazy" federal regulation barring helicopters and other motorized transport from official wilderness areas.

As Tauzin tells it, the moral of the story is that environmental laws place a higher priority on the sanctity of nature than on the well-being of humans.

But the boy's father, Robert Graham of Lake Bluff, Illinois, doesn't draw quite the same lesson. (His son survived without injury.)

When interviewed, Graham pointed out that the wilderness regulation does permit helicopter rescues if someone is in danger. Graham said officials on the scene decided that his son did not need to be airlifted.

"My problem was with that decision," Graham said. "It was the crazy idea that a 14-year-old boy lost in the mountains wasn't in danger. But that's not an argument for letting motorcycles and off-road vehicles run roughshod through the wilderness."

Carry That Weight

Yet, many of the stories heard on Capitol Hill contain an important element of truth. They reinforce the fact that as many major sources of pollution have been cleaned up, the burden of regulation is falling more heavily on people unaccustomed to the weight.

Policies that once targeted the paper, petrochemical, auto and public power industries have been making life more difficult for the dry cleaner, bakery, body shop, metal finisher and small manufacturer.

"Laws passed in the 1980s greatly increased the pervasiveness of environmental controls," Robert Sussman, a Washington lawyer and former EPA deputy administrator, wrote in an environmental journal. "As a result of Superfund enforcement, Clean Air Act (rules) and drinking water regulation, small business, cities and towns that are new to the regulatory process now confront complex and costly responsibilities and the threat of fines."

The unpopularity of these laws, along with mandatory ride-sharing and vehicle inspection rules, Sussman said, "has greatly contributed to the widespread view of EPA as a remote and heavy-handed regulator that micro-manages the daily activities of small business, local governments and individual citizens."

At the heart of much of the discontent is the belief that many environmental regulations simply are not worth the money they wind up costing industry and society.

Paul Portney, vice president of Resources for the Future, a Washington think tank that analyzes environmental policy, has estimated that urban air pollution regulations under the 1990 Clean Air Act are costing big and small businesses $20 billion annually while saving at most $12 billion in health costs.

A widely quoted critic of regulatory excess, U.S. Supreme Court Justice Stephen Breyer—a Democrat appointed by Clinton—has written that the nation is spending more money trying to eradicate the last 10 percent of pollution than it did on the first 90 percent.

In his book *Breaking the Vicious Cycle*, a critique of federal regulation, Breyer described a case in which the government sought $9 million in extra cleanup work at a toxic waste dump that already was "clean enough for children playing on the site to eat small amounts of dirt daily for 70 years without significant harm."

Republicans have proposed legislation that would make the EPA and other agencies show the benefits to society of a particular regulation justify its costs. The congressional reformers argue such a cost-benefit standard is the best way to curb the spiraling price of federal regulations, now estimated by Republicans at nearly $600 billion a year.

Calculating the Incalculable

Clinton Administration officials have balked at the proposal. Typical of the congressional reform package, they say, it goes too far.

"We have no objection to a law that says cost-benefits have to be taken into account," said EPA's Browner. "We do it all the time to help determine the best way for local government and industry to solve problems."

But the proposed law would take cost-benefit to absurd lengths, Browner argued. "We'd be trying to put a dollar value on every asthma attack prevented. Assigning monetary values to lives saved and illnesses avoided is a highly subjective process that would invite endless legal challenges."

Environmental activists, saying it is impossible to quantify many intangible benefits of regulation, argue that the move is a device to block needed protections.

"How do you calculate the benefits to society of species diversity, the joy of experiencing a wilderness or the satisfaction of knowing that humpback whales still exist out there somewhere in the deep?" said David Vladeck, a lawyer for Public Citizen, a nonprofit consumer advocacy group.

For the most part, leaders of the regulatory reform movement in Congress are careful not to present themselves as environmental killjoys. As he discussed the legislative achievements of the first 100 days of 1995 at a news conference, House Speaker Newt Gingrich cuddled a rare Southeast Asian bearcat on loan from an Ohio zoo.

But the congressional critics of federal environmental policy do not have to look any further for ammunition than the EPA itself. Although EPA officials are reluctant to shoulder much blame, they readily concede that regulatory priorities too often have been shaped by political demands and popular fears rather than by scientific analysis. . . .

Justice Breyer has written extensively about a tendency of the federal government to regulate pollutants before it can fully document the hazards they pose. He has pointed out that of 190 chemicals subject to regulation, the EPA has developed complete test data for only seven. Still, political pressure to regulate the sources of those 190 chemicals has forced the EPA to set tough standards.

The agency has required industry to use "maximum available control technology"—often meaning the most expensive technology—to reduce suspected human health hazards which, in many cases, scientists have not thoroughly evaluated. Compliance with EPA's standards has led to some costly and questionable investments.

In 1990, under orders from the EPA, officials at an Amoco Corporation oil refinery in Pennsylvania built a $41-million high-tech waste-water treatment system to capture toxic vapor before it escaped into the air. But by the time Amoco was finished with the project, testing revealed that waste water was not the main source of the vapor.

When the source was discovered, Amoco officials found that they could eliminate 90 percent of the pollution at one-quarter the cost of the waste-water plant.

The episode has become a parable of bureaucratic ineptitude fueling the uprising against federal regulation.

Acting on Basic Knowledge

No one quarrels with the idea that regulations should make sense environmentally and economically. But opponents of the pending bills say they call for a level of technical certainty that is rarely achievable. And they argue that the legislation would create a system of appeals and judicial review that will stall regulations for years.

If the proposed legislation had been in effect 20 years ago, Browner said, "we wouldn't have been able to ban lead in gasoline"—a step widely regarded as a signal achievement of environmental regulation.

"We acted on the basic knowledge of elevated lead levels in children's blood," Browner said. "We knew it was affecting children's intelligence levels. But we couldn't tell you how many IQ points would be lost or how many children would lose IQ points. And we certainly hadn't done a cost-benefit analysis."

Periodical Bibliography

The following articles have been selected to supplement the diverse views presented in this chapter. Addresses are provided for periodicals not indexed in the *Readers' Guide to Periodical Literature*, the *Alternative Press Index*, or the *Social Sciences Index*.

Patricia Adams
"Property Rights and Bioregionalism," *Cato Policy Report*, November/December 1994. Available from 1000 Massachusetts Ave. NW, Washington, DC 20001.

Marcus Colchester
"Slave and Enclave: Towards a Political Ecology of Equatorial Africa," *Ecologist*, September/October 1993.

John Bellamy Foster
"Global Ecology and the Common Good," *Monthly Review*, February 1995.

Greenpeace
"Converting to Energy for a Healthy Atmosphere," January–March 1995.

Christine E. Gudorf
"Population, Ecology, and Women," *Second Opinion*, January 1995. Available from Park Ridge Center, 211 E. Ontario, Suite 800, Chicago, IL 60611.

Yasuhiko Ishida
"Regreening the Earth: Japan's 100-Year Plan," *Futurist*, July/August 1993.

James Lis and Kenneth Chilton
"Limits of Pollution Prevention," *Society*, March/April 1993.

Farhad Mazhar
"Sustainable Development: Grassroots Initiatives in Bangladesh: Interview with Farhad Mazhar," *Multinational Monitor*, April 1993.

Steven M. Rubin, Jonathan Shatz, and Colleen Deegan
"International Conservation Finance: Using Debt Swaps and Trust Funds to Foster Conservation of Biodiversity," *Journal of Social, Political & Economic Studies*, Spring 1994.

Nira Worcman
"Trends: Boom and Bus," *Technology Review*, November/December 1993.

Kitazawa Yoko
"The Politics of Sustainable Development," *Connexions*, no. 41, 1993.

John E. Young
"Global Network: Computers in a Sustainable Society," *Worldwatch Paper*, no. 115, September 1993. Available from 1776 Massachusetts Ave. NW, Washington, DC 20036.

For Further Discussion

Chapter 1

1. What do Al Gore and Julian L. Simon choose as measures of the environmental health of the planet? How do their choices of measures differ, and which do you find most meaningful? Explain why.

2. Gore, Simon, Barbara Ruben, and Stephen Budiansky all include the arguments of their opponents in their viewpoints. Which author do you believe uses this strategy most effectively? Why? How does this author handle the opponent's arguments differently from the others?

3. Budiansky argues that scientists brought an environmental backlash upon themselves by exaggerating environmental threats. Ruben argues that this backlash was initiated by industries that hired their own scientists to produce research endorsing their environmentally unsound business practices. What are the credentials of the sources these authors quote? Are their ideas being quoted fully enough? Why or why not?

Chapter 2

1. K.H. Jones and Jonathan Adler argue that air quality in the United States (outside of Southern California) is not a problem, while the United Nations Environment Programme takes the position that air pollution is a significant threat. On what information does each party base its standard for safe air? Whose information do you consider more relevant, and why?

2. Both Melissa Healy and Jonathan Tolman use the 1993 cryptosporidium outbreak in Milwaukee to support their arguments about the safety of America's drinking water. What assumptions do you think led each author to draw such divergent conclusions from the same incident?

Chapter 3

1. Alan Thein Durning asserts that the consumer society produced by industrialism has impoverished Americans both spiritually and environmentally; Stephanie Moussalli argues that subordinating consumer demand to environmental concerns is hostile to human welfare. What do these assertions reveal about each author's definition of human welfare?

2. Neal Peirce argues that American suburbia is responsible for the urban blight and social deterioration of inner cities. Randal O'Toole insists that inner cities, and in particular their design, are the source of these problems. Based on your reading of these viewpoints, what would you propose as the most socially and environmentally beneficial approach to developing cities and neighborhoods?

3. F. Herbert Bormann states that lawn pesticides may pose a threat to human health; Leonard T. Flynn argues that a lack of data about the adverse effects of pesticides is ample evidence of their harmlessness. In your opinion, is the information Flynn presents sufficient to disprove Bormann's statement? Why or why not?

Chapter 4

1. Virginia Warner Brodine and James Owen Rice agree that forest preservation need not cost loggers their jobs, yet Brodine supports logging bans on old growth forests while Rice does not. How are their differing views related to their fundamental differences of opinion on issues of environmentalism and capitalism?

2. Fred L. Smith Jr. argues that innovation is a more important consideration in predicting the future of humankind than is concern over the limits of natural resources. John Miller argues that the planet's resources are finite and that humans' survival depends on their willingness to recognize this. How does each author support his argument? Whose argument seems stronger, and why?

Chapter 5

1. Bennett C. Karp believes that value is created by humans when they expend the effort necessary to extract and modify natural resources in the production of goods. Paul Hawken sees value as extending beyond the cost of production to include the environmental impact of that production and resulting waste disposal. What do these contrasting definitions reveal about the authors' attitudes towards the natural world and humans' place within it? Whose attitude most closely matches your own, and why?

2. Frank Clifford quotes EPA chief administrator Carol Browner as saying that the level of technical certainty called for by opponents of federal environmental regulation is both unreasonable and unattainable. Does the material Clifford presents affect your assessment of Fife Symington's proposal to "scrap" the existing regulatory structure? Why or why not?

Organizations to Contact

The editors have compiled the following list of organizations concerned with the issues debated in this book. The descriptions are derived from materials provided by the organizations. All have publications or information available for interested readers. The list was compiled on the date of publication of the present volume; names, addresses, and phone numbers, fax numbers, and e-mail addresses may change. Be aware that many organizations take several weeks or longer to respond to inquiries, so allow as much time as possible.

American Crop Protection Association (ACPA)
1156 15th St. NW, Suite 400
Washington, DC 20005
(202) 296-1585
fax: (202) 463-0474

ACPA is an association of firms that produce agricultural chemical products like herbicides, pesticides, defoliants, and soil disinfectants. It contains legislative and regulatory departments and maintains committees on environmental management, public health, and toxicology. The association promotes the use of chemicals in farm production. It publishes the periodic *Bulletin* and the quarterly *Growing Possibilities*.

Canadian Forestry Association (CFA)
185 Somerset St. West, Suite 203
Ottawa, ON K2P 0J2
CANADA
(613) 232-1815
fax: (613) 232-4210

CFA works for an improved forest management that would satisfy the economic, social, and environmental demands on Canadian forests. It explores conflicting perspectives on forestry-related topics in its biannual *Forest Forum*.

Competitive Enterprise Institute (CEI)
1001 Connecticut Ave. NW, Suite 1250
Washington, DC 20036
(202) 331-1010
fax: (202) 331-0640

CEI encourages the use of private incentives and property rights to protect the environment. It advocates removing government barriers to establish a system in which the private sector would be responsible for the environment. CEI's publications include the monthly newsletter *CEI Update* and the monographs *Federal Agricultural Policy: A Harvest of Environmental Abuse* and *Conserving Biodiversity: Resources for Our Future*.

Earth Island Institute
300 Broadway, Suite 28
San Francisco, CA 94133
(415) 788-3666
fax: (415) 788-7324

Earth Island Institute is a nonprofit organization that focuses on environmental issues and their relation to such concerns as human rights and economic development in the Third World. The institute publishes the quarterly *Earth Island Journal: An International News Magazine.*

Environmental Defense Fund (EDF)
257 Park Ave. South
New York, NY 10010
(212) 505-2100
fax: (212) 505-0892

EDF is a public interest organization that is dedicated to the protection and improvement of environmental quality and public health. It advocates reform of public policy in the areas of global atmosphere programs, solid-waste issues, air quality, energy, water resources, agriculture, and international environment. The fund's publications include *Biotechnology's Bitter Harvest, Developing Policies for Responding to Climatic Change, Chlorofluorocarbon Policy,* and the bimonthly newsletter *EDF Letter.*

Foundation for Research on Economics and the Environment (FREE)
502 S. 19th Ave.
Bozeman, MT 59715
(406) 585-1776

FREE is a research and education foundation committed to freedom, environmental quality, and economic progress. It works to reform environmental policy by using the principles of private property rights, the free market, and the rule of law. FREE publishes the quarterly newsletter *FREE Perspectives on Economics and the Environment* and produces a biweekly syndicated op-ed column.

Friends of the Earth
1025 Vermont Ave. NW, 3rd Fl.
Washington, DC 20005
(202) 783-7400
fax: (202) 783-0444

Friends of the Earth is dedicated to protecting the earth from environmental disaster and to preserving biological and ethnic diversity. The organization encourages ozone and groundwater protection, toxic waste cleanup, reform of the World Bank, and the use of tax dollars to protect the environment. It publishes the bimonthly newsletter *Friends of the Earth* and the books *Crude Awakening, the Oil Mess in America: Wasting Energy, Jobs, and the Environment* and *Earth Budget: Making Our Tax Dollars Work for the Environment.*

Greenpeace
1436 U St. NW
Washington, DC 20009
(202) 462-1177
fax: (202) 462-4507

Greenpeace opposes nuclear energy and the use of toxics and supports ocean and wildlife preservation. It uses controversial direct-action techniques and strives for media coverage of its actions in an effort to educate the public. It publishes the quarterly magazine *Greenpeace* and the books *Radiation and Health, Coastline,* and *The Greenpeace Book on Antarctica.*

Minnesota Mining and Manufacturing (3M)
Environmental Communications
Bldg. 225-3S-05
St. Paul, MN 55144-1000
(612) 733-1135
fax: (612) 733-6557

3M has implemented various toxic-waste reduction measures that have saved the company several hundred million dollars. Numerous environmental leaders have singled out 3M as a positive example of corporate environmental responsibility. The company makes available an information packet on its "Pollution Prevention Pays" program and an "Environmental Progress Report" showing 3M's environmental goals and the strategies the company uses to achieve them.

National Solid Wastes Management Association (NSWMA)
4301 Connecticut Ave. NW, Suite 300
Washington, DC 20008
(202) 244-4700

NSWMA is a trade organization of industries involved in garbage collection, recycling, landfills, and treatment and disposal of hazardous and medical wastes. It lobbies for laws that are environmentally sound but that still allow communities to dispose of their waste. It publishes the monthly magazine *Waste Age: The Authoritative Voice of Waste Systems and Technology* and the biweekly newsletter *Recycling Times.*

The Political Economy Research Center (PERC)
502 S. 19th Ave., Suite 211
Bozeman, MT 59715
(406) 587-9591
fax: (406) 587-7555

PERC is a research center that provides solutions to environmental problems based on free market principles and the importance of private property rights. Its publications include the quarterly newsletter *PERC Report* and papers in the PERC Policy Series dealing with environmental issues.

Reason Foundation
3415 S. Sepulveda Blvd., Suite 400
Los Angeles, CA 90034
(310) 391-2245
fax: (310) 391-4395

The Reason Foundation is a national public policy research organization. It specializes in a variety of policy areas, including the environment, education, and privatization. The foundation publishes the monthly magazine *Reason* and the books *The Case Against Electric Vehicle Mandates in California*, *Solid Waste Recycling Costs—Issues and Answers*, and *Global Warming: The Greenhouse, White House, and Poorhouse Effects*.

Stockholm Environment Institute (SEI)
11 Arlington St.
Boston, MA 02116-3411
(617) 266-8090
fax: (617) 266-8303
e-mail: postmaster@tellus.com
web site: http://www.channel1.com/users/tellus/seib.html

Headquartered in Sweden, SEI is a research institute that operates through an international network. The institute focuses on a variety of environmental issues, including energy use, freshwater resources, and climate change. SEI publications include the quarterly newsletters *SEI: An International Environment Bulletin* and *Renewable Energy for Development*.

Together Foundation for Global Unity
130 S. Willard St.
Burlington, VT 05401
(802) 862-2030
fax: (802) 862-1890

The Together Foundation is a nonprofit organization that provides information on ecological, environmental, sustainable growth, and human rights issues. It publishes the quarterly newsletter *Together News* and provides the electronic bulletin board TogetherNet.

Union of Concerned Scientists (UCS)
PO Box 9105
Cambridge, MA 02238-9105
(617) 547-5552
fax: (617) 864-9405
e-mail: menu@ucsusa.org

UCS works to advance responsible public policies in areas where science and technology play a vital role. Its programs focus on safe and renewable energy technologies, transportation reform, arms control, and sustainable agriculture. UCS publications include the quarterly magazine *Nucleus* and the briefing papers *Alternative Transportation Fuels*, *Environmental Impacts of Renewable Energy Technologies*, and *The Global Environmental Crisis: Causes, Connections, and Solutions*.

Bibliography of Books

Douglas Adams — *Last Chance to See.* New York: Harmony Books, 1991.

Terry Anderson and Donald Leal — *Free Market Environmentalism.* Boulder, CO: Westview Press, 1991.

Joseph L. Bast, Peter J. Hill, and Richard C. Rue — *Eco-Sanity: A Common-Sense Guide to Environmentalism.* Lanham, MD: Madison Books, 1994.

A.S. Bhalla, ed. — *Environment, Employment, and Development.* Geneva, Switzerland: International Labour Office, 1992.

David Ross Brower with Steve Chapple — *Let the Mountains Talk, Let the Rivers Run: A Call to Those Who Would Save the Earth.* 1st ed. San Francisco: HarperCollins West, 1995.

Daniel D. Chiras — *Beyond the Fray: Reshaping America's Environmental Response.* Boulder, CO: Johnson Books, 1990.

Commission on Geosciences, Environment, and Resources — *Restoration of Aquatic Ecosystems: Science, Technology, and Public Policy.* Washington, DC: National Academy Press, 1992.

Committee on Wastewater Management for Coastal Urban Areas, National Research Council — *Managing Wastewater in Coastal Urban Areas.* Washington, DC: National Academy Press, 1993.

David Edward Cooper — *The Environment in Question: Ethics and Global Issues.* New York: Routledge, 1992.

John DeWitt — *Civic Environmentalism: Alternatives to Regulation in States and Communities.* Washington, DC: CQ Press, 1994.

Irene Diamond — *Fertile Ground: Women, Earth, and the Limits of Control.* Boston: Beacon Press, 1994.

Richard Douthwaite — *The Growth Illusion: How Economic Growth Has Enriched the Few, Impoverished the Many, and Endangered the Planet.* Tulsa, OK: Council Oaks Books, 1993.

Roger C. Dower and Mary Beth Zimmerman — *The Right Climate for Carbon Taxes: Creating Economic Incentives to Protect the Atmosphere.* Washington, DC: World Resources Institute, 1992.

Peter Ferdinand Drucker — *The Ecological Vision: Reflections on the American Condition.* New Brunswick, NJ: Transaction Publishers, 1993.

Gordon K. Durnil	*The Making of a Conservative Environmentalist: With Reflections on Government, Industry, Scientists, the Media, Education.* Bloomington: Indiana University Press, 1995.
William O. Dwyer and Frank C. Leeming	*Earth's Eleventh Hour: Environmental Readings.* Boston: Allyn and Bacon, 1995.
Gregg Easterbrook	*A Moment on the Earth: The Coming Age of Environmental Optimism.* New York: Viking, 1995.
Jim Eggert	*Meadowlark Economics: Perspectives on Ecology, Work, and Learning.* Armonk, NY: M.E. Sharpe, 1992.
David Ehrenfeld	*Beginning Again: People and Nature in the New Millennium.* New York: Oxford University Press, 1993.
Paul Faeth	*Growing Green: Enhancing Environmental and Economic Performance in U.S. Agriculture.* Washington, DC: World Resources Institute, 1995.
Michael Fumento	*Science Under Siege: Balancing Technology and the Environment.* 1st. ed. New York: Morrow, 1993.
Gail Enid Gelburd	*Creative Solutions to Ecological Issues.* New York: Council for Creative Projects, 1993.
Eli Gifford and R. Michael Cook, eds.	*How Can One Sell the Air? Chief Seattle's Vision.* Summertown, TN: Book Publishing Company, 1992.
John Holtzclaw	*Using Residential Patterns and Transit to Decrease Auto Dependence and Costs.* New York: Natural Resources Defense Council, 1994.
Michael F. Jacobson	*Marketing Madness: A Survival Guide for a Consumer Society.* Boulder, CO: Westview Press, 1995.
Jimmie M. Killingsworth and Jacqueline S. Palmer	*Ecospeak: Rhetoric and Environmental Politics in America.* Carbondale: Southern Illinois University Press, 1992.
G.B. Leslie and F.W. Lunau, eds.	*Indoor Air Pollution: Problems and Priorities.* New York: Cambridge University Press, 1992.
Gene E. Likens	*The Ecosystem Approach: Its Use and Abuse.* Oldendorf/Luhe, Germany: Ecology Institute, 1992.
Anil Markandya and Julie Richardson, eds.	*Environmental Economics: A Reader.* New York: St. Martin's Press, 1993.
David Maybury-Lewis	*Millennium: Tribal Wisdom and the Modern World.* New York: Viking, 1992.

209

Index